Principles, Application and Assessment in Soil Science

Editor

Stefanos Palmer

scitus academics

Principles, Application and Assessment in Soil Science
Edited by **Stefanos Palmer**

ISBN: 978-1-68117-322-1
Library of Congress Control Number: 2016937632

SCITUS Academics LLC,
www.scitusacademics.com
Box No. 4766, 616 Corporate Way,
Suite 2, Valley Cottage,
NY 10989

Printed in United States of America on Acid Free Paper

Preface

Our soil dependence and interest in it is important to investigate changes in soil processes. The diversity and dynamics of the soil help us to understand the differences in the soil processes. This allows us to take the measures necessary to protect the soil and thus to promote the health of the soil. Soil is a natural body composed of solids (minerals and organic matter), liquids and gasses that occur on the surface of the land, occupy space and are characterized by one or both: horizons or layers that can be distinguished from the original material by adding, losing, transferring and transforming energy and matter or by the ability to support rooted plants in the soil science. The study of man's effects on the ground has been there for a long time. From basic agricultural concepts to crop rotation to modern laboratory-mixed soils and fertilizers, all these ideas stem from the study of the soil and its use. It wasn't until the 20th century, however, that the field of soil science became a recognized scientific area. While many soil science uses are well known, some are even less evident to a layperson. These areas have a major role to play in remediating soil contamination from landfills, toxic dumping and environmental accidents.

This book gives an up to date account of the current knowledge of recent practices and evaluations in soil science. It also presents an inclusive assessment of the impact of the application of residues/waste on the soil properties and the plant adaptation and growth mechanisms.

Table of Contents

CHAPTER 1

Soil Nutrients, Landscape Age, and Sphagno-Eriophoretum vaginati Plant Communities in Arctic Moist-Acidic Tundra Landscapes

Joel A. Mercado-Díaz*, William A. Gould, Grizelle González

USDA Forest Service, International Institute of Tropical Forestry, Jardín Botánico Sur, 1201 Calle Ceiba, Río Piedras, Puerto Rico

ABSTRACT

Most research exploring the relationship between soil chemistry and vegetation in Alaskan Arctic tundra landscapes has focused on describing differences in soil elemental concentrations (e.g. C, N and P) of areas with contrasting vegetation types or landscape age. In this work we assess the effect of landscape age on physico-chemical parameters in organic and mineral soils from two long- term research sites in northern Alaska, the Toolik Lake and Imnavait grids. These two sites have contrasting landscape age but similar vegetation composition. We also used correlation analysis to evaluate if differences in any of these parameters were linked with between-site variation in the abundance of growth forms. Our analysis was narrowed to soils in Sphagno-

Eriophoretum vaginati plant communities. We found no significant differences between these sites for most parameters evaluated, except for total Ca which was significantly higher in organic soils from Imnavait vs. Toolik and total Na which was significantly higher in mineral horizons from Toolik compared to Imnavait. Moreover, the abundance of non-Sphagnum mosses was positively correlated with total Ca in organic soils, whereas the abundance of forbs, non-Sphagnum mosses and bryophytes was negatively correlated with total Na in mineral soils. We suggest that differences in the concentration of these two elements are most likely tied to landscape age differences between these sites. However, since observed dissimilarity in terms of total Ca in organic soils and total Na in mineral soils is concordant with correlation patterns observed between these elements and the aforementioned growth forms, it is likely that existing differences in vegetation composition between these sites are also influencing the concentration of these elements in soils, particularly that of Ca, since non-Sphagnum mosses are dominant above organic soils and are therefore expected to significantly influence biogeochemical processes at this horizon. Thus, we conclude that except for organic Ca and mineral Na, there is little difference between these sites in terms of their soil physico-chemical properties. We suggest that most of the influence of landscape age on evaluated parameters is masked by factors such as moderate cryoturbation and similarities in terms of vegetation properties and climate. These observations are relevant as they suggest a linkage between soil chemistry and vegetation composition in this tundra region.

KEYWORDS

Arctic Alaska, Soil Nutrients, Moist-Acidic Tussock Tundra, Vegetation

1. INTRODUCTION

Studies have shown that landscape age (time since deglaciation) is an important factor influencing essential biogeochemical processes in Arctic tundra soils [1] -[6] . For instance, Hobbie and Gough [7] compared foliar and soil nutrients between landscapes that were deglaciated > 50,000 (moist-acidic tundra [MAT]) and less than 11,500 years ago (moist non-acidic tundra [MNT]) and found higher rates of net N mineralization, cation exchange capacity and exchangeable base cations in soils at the geologically older site. Contrasting landscape age has also been proposed as a potential factor explaining variation in litter decomposition rates [8] and the rate of processes like C and N cycling in tundra soils [9] .

In the Alaskan Arctic, differences in landscape age are also intimately related to differences in soil acidity [2] [3] . A number of important biological attributes are influenced by gradients of soil pH in this region. For example, some studies have documented the relationship between contrasting vegetation types with a distinct pH boundary that separates MAT and MNT [7] [9] . Considerable variation in species and growth form dominance within some tussock tundra communities has been associated with variation in soil pH [10] [11] . Differences in soil pH can also affect specific plant community attributes like vascular-plant species richness [12] . More recently, Eskelinen et al. [13] proposed an indirect effect of soil pH on vegetation via the evolution of bacteria- based microbial communities in alkaline soils where the properties of forb-produced organic matter were possibly sustaining the prevalence of soil bacteria.

Vegetation can also influence important soil processes like nutrient cycling which regulate nutrient concentrations and the size of C and N pools in tundra soils [14] . Since vegetation effects on soils are thought to be primarily related to the accumulation of organic material and nutrients [15], variation in specific traits influenced by plant communities, such as litter chemistry, is believed to play an important role in some of these processes. For instance, altered litter quality resulting from changes in species composition is known to affect processes like soil N mineralization [16]. However, some evidence appears to indicate that other landscape-scale soil processes are less likely to be significantly affected by vegetation. For example, Hobbie and Gough [8] demonstrated that variation in plant species composition did not account for differences in litter decomposition between MAT and MNT.

The linkage between patterns in vegetation composition and soils in tundra landscapes has not been characterized thoroughly, although several studies have contributed significantly in this direction. Chu and Grogan [17] found that variation in total C and N, Dissolved Organic Carbon (DOC) and N (DON), mineral N and N mineralization potential in organic soils in Daring Lake, Canada has been directly related to differences in vegetation types. Likewise, Eskelinen et al. [13] observed that forb-rich non-acidic heaths were associated with low C:N and low soluble N:phenolics ratios in soils, whereas shrub-dominated acidic heaths were associated with high values of these ratios. Understanding the reciprocal relationship between vegetation and soil dynamics in Arctic tundra ecosystems is challenging mostly because vegetation is ultimately controlled by meso-topographic relationships (slope position and soil moisture), micro-scale disturbances and factors related to long-term landscape evolution [4] [18] . This suggests that plant communities exhibit considerable spatial variability along soil chemical gradients.

In this study we compare a number of soil physico-chemical parameters from organic and mineral horizons of two study sites in northern Alaska that differ in landscape age, i.e., time since deglaciation 55 k and 125 k years, but have similar vegetation composition [18]. These are long- term vegetation monitoring areas which have been intensively used since the late 1980s to study the effects of climate change on tundra vegetation [19] -[21] . Our analysis is focused on soils derived from areas dominated by Sphagno-Eriophoretum vaginati plant communities' sensu Walker et al. [18] , the most ubiquitous plant community within these sites. Little information on soil chemistry has been published from soils within these long-term monitoring sites. Only recently the works of Whittinghill and Hobbie [5] [6] and Keller et al. [23] have shown that the two landscape ages represented by these two study sites are similar in terms of several soil chemical parameters. The present work aims to advance the understanding of soil processes in this region by including analyses on physico-chemical properties of both mineral and organic horizons and other aspects of soil chemistry that are not necessarily evaluated in detail by those studies.

We also assess between-site differences in the abundance of a number of growth forms and subsequently evaluate the relationship between those differences and the variation between sites in terms of particular physico-chemical parameters to make inferences regarding how vegetation composition may be affecting soil physico-chemical properties at these two sites, and vice versa. These observations will contribute to a better understanding on how vegetation composition, landscape age and soil chemistry are reciprocally linked in this region.

2. MATERIALS AND METHODS

2.1. Study Area

This study was carried out in two 1 km^2 research grids, Toolik Lake and Imnavait Creek, in the vicinity of the Toolik Lake Field Station, Alaska, located north of the Brooks Range in the Southern Foothills Physiographic Province of the Alaskan North Slope [24] [25] (Figure 1). The region is underlain by continuous permafrost which is 250 - 300 m thick [26] . Both grids are dominated by MAT vegetation; however their landscapes are slightly different due primarily to differences in glacial age [18] [27] [28] . The Imnavait grid is in the headwaters of the Imnavait Creek, a small beaded tributary of the Kuparuk River basin [18] . This site lies on the Sagavanirktok (Middle Pleistocene) glacial drift which deglaciated about 125,000 years ago [27] [28] . Topography is dominated by gently rolling hills and elevation within the watershed varies from about 770 to 980 m [22] . Most of the Toolik Lake grid lies in a younger substrate (Itkillik I glacial drift) which deglaciated during the late Pleistocene (ca. 55,000 years), but includes areas of Itkillik II outwash towards the east [27] [28] . The landscape at this site is more heterogeneous than at Imnavait and is dominated by small glacial lakes, kames and moraines. Elevations range from 670 to 850 m [18].

Both grids share similar meteorological conditions due to their close proximity (<12 km). Mean annual surface atmospheric temperature (SAT) from 1989-2008 have been −8.5°C; whereas average annual precipitation during the same period was 312 mm [29] . Linear trend analysis of mean annual SAT and mean annual precipitation revealed no trends in these parameters suggesting that they have remained stable over the last two decades [29] .

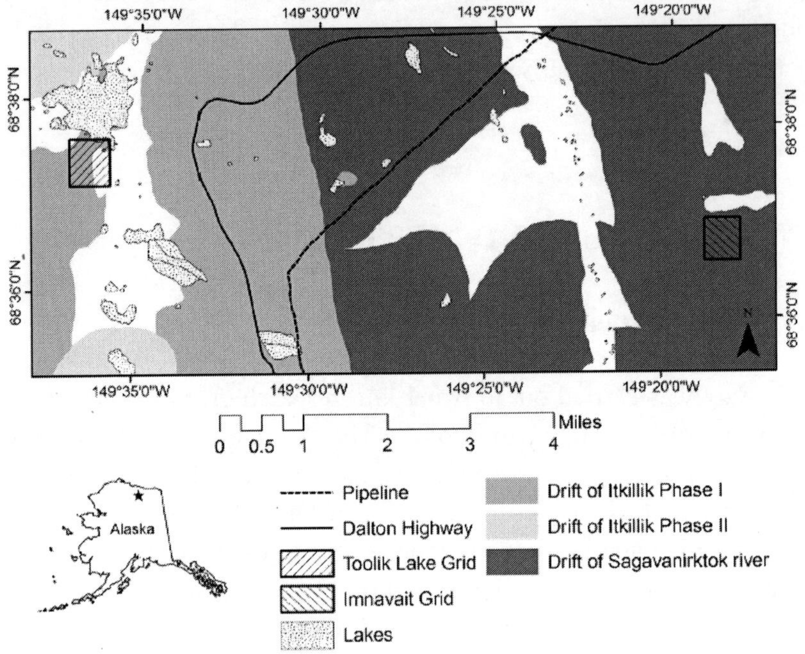

Figure 1. Landscape ages and the location of the Toolik Lake and Imnavait Creek 1 km^2 grids in the Upper Kuparuk River region in Northern Alaska.

2.2. Study Design

There are 157 permanent 1m^2 vegetation plots within these grids, 72 at Imnavait and 85 at Toolik. Plots are located equidistantly at 100 meters from each other within the grids [19] . These are non-manipulative plots that are currently being studied for analyzing long term changes in vegetation composition and structure in this region [30]. We limited our analysis to a subset of these plots that are established in areas with Sphagno-Erio- phoretum vaginati plant communities. This plant community has a broad spatial extent which allows us to extrapolate results to a landscape-scale level. Likewise, confounding effects that may result from grid-scale variation in plant communities could be reduced by focusing on a single plant community. The Sphagno-Eriophoretum vaginati plant community is the most dominant plant community within the grids and is considered the zonal vegetation of mesic slopes throughout the Arctic Foothills [18] . It typically occurs on ice-rich sediments with shallow active layers and low soil pH [31] . The most conspicuous plant species is the tussock-forming sedge Eriophorum vaginatum which dominates particularly in stable hillslope shoulders and upper backslopes; whereas shrubs like Betula nana and Salix pulchra tend to become dominant on footslopes and associate with deep Sphagnum spp. mats, other mosses like Aulacomnium turgidum, Hylocomnium splendens and lichens like Peltigera aphtosa and Cladonia spp. [18] [22] [32] . Sphagno-Eriophoretum vaginati plant communities' sensu Walker et al. [18] are contained within the Moist-tussock sedge, dwarf shrub, moss tundra physiognomic unit [24] and mostly coincide with Eriophorum vaginatum-Sphagnum spp. plant communities' sensu Walker et al. [22] .

Plant communities within the Toolik and Imnavait grids were identified using geographical layers prepared by Walker [33] and published maps of the region [24] . There are 33 vegetation types represented in the Upper Kuparuk river region [22] [31] . Walker et al. [18] classified vegetation of both Toolik and Imnavait grids into five associations and 15 community types. We performed an overlay analysis using a geographic information system (GIS) and selected 24 plots (12 at Toolik and 12 at Imnavait) representing Sphagno-Eriophoretum vaginati plant communities. We extracted growth form abundance data from selected plots from a long-term plant community dataset. This data corresponded to vegetation sampling realized in 2007 (Imnavait) and 2008 (Toolik) [34].

2.3. Sampling and Nutrient Analysis

Soil samples were collected 1 - 2 meters apart from selected plots in areas with visually similar vegetation composition. Samples of organic and mineral soils were collected near each plot following the procedures described below. Mineral soils near two of the 24 plots (both at Toolik) were frozen at the time of collection and not sampled. Soil sampling was realized during August, 2008.

A shovel was used to create a 25 × 25 cm pit to collect soil samples in each selected plot. Pits reached an approximate depth of 20 cm, or deeper until the upper 5 - 7 cm portion of the mineral horizon was exposed. Three samples of mineral soils (n = 3) were taken at each pit by pushing a stainless steel soil core with plastic core inserts (aprox. 98 cc.) horizontally into the soil profile. These samples were used individually to calculate soil moisture, bulk density and elemental concentrations in this horizon. Mineral soil moisture was calculated after oven drying samples at 105°C for 48 hours in laboratory facilities at Toolik Lake Field Station. Depending on soil conditions, each organic soil sample (n = 1) was extracted horizontally at the soil profile or vertically using a small bread knife and were then placed into labeled cloth bags. When samples were collected vertically, we removed the upper layer of soil where live material was evident and the lower layer where the organic layer diffuses into the mineral soil layer. Each organic soil sample had approximately twice the volume of mineral soil samples. All samples (except those used for estimating mineral soil moisture) were sent to the International Institute of Tropical Forestry Soils Laboratory (USDA-USFS) in San Juan, Puerto Rico and were processed within 8 days. Mineral soil samples for elemental concentration and bulk density determinations were oven dried at 40°C for two weeks, whereas those from organic soils were air dried over the same period. We used a Foss Tecator Cyclotec (model 1093) mill to ground the samples and then passed them through a 1 mm stainless steel sieve. Roots of considerable size and other live material were excluded from the samples.

Total Nitrogen (N) and Total Carbon (C) were analyzed using the dry combustion method by means of a LECO TruSpec CN Analyzer [35] . The procedure used is a modified version of the Organic Application Note titled "Carbon and Nitrogen in Soil and Sediment" obtained from LECO Corp. [36] . The dry combustion method was also used to determine Total Sulfur (S), utilizing the LECO TruSpec (Add-On Module) S Analyzer [37] . The procedure used is from LECO [38] and is titled "Sulfur in Cement, Fly Ash, Limestone, Soil, and Ore". In the dry combustion method a small weighted sample is combusted in a high temperature furnace (950°C for the LECO TruSpec CN Analyzer and 1450°C for the LECO TruSpec S Analyzer) and in a stream of purified oxygen. Total Carbon is measured as CO_2 by an infrared detector and

Total Nitrogen is determined as N_2 by a thermal conductivity cell detector. Total Sulfur as SO_2 is also detected by an infrared.

Ground material was analyzed for elemental concentrations of Phosphorus (P), Aluminum (Al), Potassium (K), Calcium (Ca), Magnesium (Mg), Manganese (Mn), Sodium (Na) and Iron (Fe). Soil samples were digested with concentrated HNO_3, 30% H_2O_2 and concentrated HCl using a modified version of the method recommended by Luh Huang and Schulte [39] and analyzed by means of a Spectro Plasma Emissions Spectrometer (Spectro Ciros ICP).

Subsamples were oven dried at 105°C for 24 hrs. A moisture factor was calculated and applied to each analysis [40] . These same subsamples were later ignited at 490°C in a muffle furnace (for at least 8 hrs.) to determine Loss-on-Ignition (LOI). Soil pH was measured in a 1:1 (soil:water) solution using an Orion Ionanalyzer Model 901 with a combination pH electrode [41] .

Soil moisture of mineral and organic soil samples was determined by calculating the percentage weight loss after drying. Organic horizon thickness was averaged for each plot and was calculated using three different measurements of the organic horizon width at randomly selected areas of each pit.

2.4. Statistical Analysis

We analyzed and treated as independent variables each physico-chemical parameter and the abundance of particular growth forms. We used Exploratory Data Analysis (EDA) to corroborate if these variables conformed to parametric testing assumptions. We assessed normality using the Shapiro-Wilk test along with Normal Q-Q plots. Except for "% Moisture" which was analyzed using non-parametric Mann-Whitney U-test, results from EDA supported the use of Independent Samples T-Test for most variables. Variables that initially failed EDA tests were mostly affected by few extreme outliers and behave normally after their exclusion.

We used Pearson's correlation coefficient (r) to evaluate how physico-chemical parameters that were significantly different between sites relate to growth forms that were also found to have significantly different abundances at each site. We combined data from both sites that corresponded to these two variables. This resulted in 22 data points for correlation analysis. Correlation tests were performed after absolute abundances values in the dataset were converted to percent cover of growth forms at each plot.

3. RESULTS

We found no significant differences between Imnavait and Toolik in terms of the thickness of the organic horizon and the bulk density of the mineral horizon (Table 1). Soils at both sites were similarly acidic whereas comparable LOI values are indicative of similar organic matter content. Only percent moisture was significantly higher in mineral soils from Imnavait vs. Toolik. Nonetheless, higher moisture recorded at Imnavait is possibly linked to higher precipitation activity that was observed at this site during sampling.

There were no differences in the concentration of C, N and S between both sites for both organic and mineral horizons (Figure 2(a), Figure 2(b) and Table 1). There were also no significant differences between sites for the concentration of other elements, except for Ca which was significantly higher in organic horizons of the older Imnavait site (M = 5.56, SD = 4.21) compared to the younger Toolik site (M = 2.74, SD = 1.97), $t(21) = 2.09$, $p < 0.05$ (Figure 3(a)) and Na which was significantly higher in mineral soils of Toolik (M = 0.27, SD = 0.07) compared to Imnavait (M = 0.17, SD = 0.10), $t(20) = 2.75$, $p < 0.05$ (Figure 3(b) and Table 1).

We found significant between-site differences in the abundances of several growth forms (Table 2and Figure 4). The abundance of shrubs was significantly higher at Toolik (M = 32.25, SD = 5.36) vs. Imnavait (M = 26.51, SD = 7.04); $t(22) = -2.25$, $p < 0.05$. In contrast, forbs were significantly more abundant at Imnavait (M = 5.99, SD = 5.52) vs. Toolik (M = 1.60, SD = 1.69); $t(22) = 2.64$, $p < 0.05$. Similarly, non-

Sphagnum mosses at Imnavait (M = 27.42, SD = 4.92) were more abundant than at Toolik (M = 10.60, SD = 5.31); t (22) = 8.05, $p < 0.00$, and bryophytes more abundant at Imnavait (M = 38.50, SD = 5.07) than at Toolik (M = 24.99, SD = 9.45); t (22) = 4.36, $p < 0.05$.

Table 1. Physical and chemical properties of mineral and organic soils from the Sphagno-Eriophoretum vaginati plant communities in Toolik Lake and Imnavait Creek grids. Between-site comparisons were assessed with Independent Samples T-Test unless otherwise noted. Sample size within parentheses unless otherwise indicated with superscripts. SD = Standard deviation. Statistical significance achieved at $p < 0.05$.

Variables	Mineral					Organic				
	Imnavait (n = 12)		Toolik (n = 10)		p-values	Imnavait (n = 12)		Toolik (n = 12)		p-values
	Mean	SD	Mean	SD		Mean	SD	Mean	SD	
Total elements										
C (%)	3.41[a]	2.13	4.01[c]	1.84	ns	34.85	7.36	29.15	10.74	ns
N (%)	0.14[a]	0.12	0.15[c]	0.10	ns	1.55	0.34	1.24	0.47	ns
S (%)	0.02[a]	0.01	0.02[c]	0.01	ns	0.19	0.05	0.17	0.08	ns
Al (mg/g)	12.18[a]	3.55	11.88	2.75	ns	8.57	4.78	8.70	3.09	ns
P (mg/g)	0.45	0.19	0.45[c]	0.18	ns	1.51	0.44	1.37	0.67	ns
Na (mg/g)	0.17[b]	0.10	0.27	0.07	$p < 0.05$	1.49	0.87	1.26	0.55	ns
Mn (mg/g)	0.30[a]	0.20	0.26	0.20	ns	6.70	7.80	4.81[a]	7.48	ns
Ca (mg/g)	0.79[a]	0.30	0.86	0.37	ns	5.56[a]	4.21	2.74	1.97	$p < 0.05$
Fe (mg/g)	27.20	7.82	27.51	8.26	ns	25.36[a]	12.64	22.32	12.27	ns
Mg (mg/g)	2.48[a]	0.84	2.14	0.28	ns	1.38[a]	0.33	1.14	0.53	ns
K (mg/g)	0.73[a]	0.27	0.79	0.20	ns	0.84[a]	0.22	0.78	0.26	ns
Organic horizon thickness (cm; n = 3)	-	-	-	-	-	13.04	1.22	10.63[c]	1.88	ns
Bulk density (g/cc)	1.25	0.14	1.46	0.12	ns	-	-	-	-	-
% Moisture	30.34	19.02	12.65	15.84	$p < 0.05$[*]	-	-	-	-	-
LOI (%)	10.52	6.18	9.88[c]	3.48	ns	68.64	12.95	58.23	19.74	ns
pH (H_2O)	4.53	0.12	4.41	0.19	ns	4.80	0.72	4.34	0.53	ns

[a]n = 11; [b]n = 10; [c]n = 9; [*]Mann-Whitney *U*-test.

Because Ca in organic soils and Na in mineral soils were the only elements exhibiting a statistically significant difference between sites, correlation analysis was focused on evaluating the relationship of these two elements with the abundance those growth forms found to have contrasting abundances between sites. In this sense, we found the abundance of non-Sphagnum mosses to be positively correlated with the total concentration of Ca in organic soils ($r = 0.48$, $n = 24$, $p < 0.05$). We also found that the total concentration of Na in mineral soils was negatively correlated with the abundances of forbs ($r = -0.486$, $n = 22$, $p < 0.05$), non-Sphagnum mosses ($r = -0.470$, $n = 22$, $p < 0.05$) and bryophytes ($r = -578$, $n = 22$, $p < 0.01$).

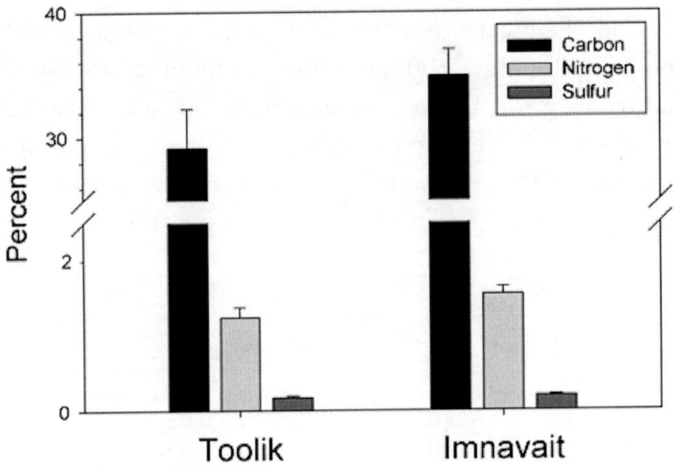

(a)

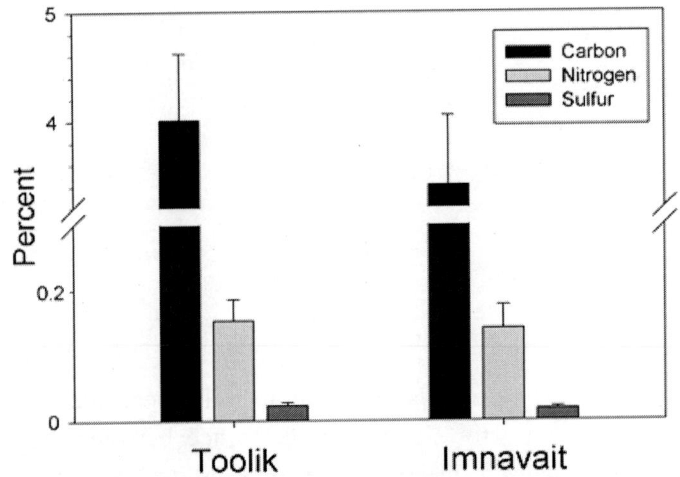

(b)

Figure 2. Differences between Toolik Lake and Imnavait Creek grids in terms of mean total concentration (%) of C, N and S in (a) organic horizons and (b) mineral horizons.

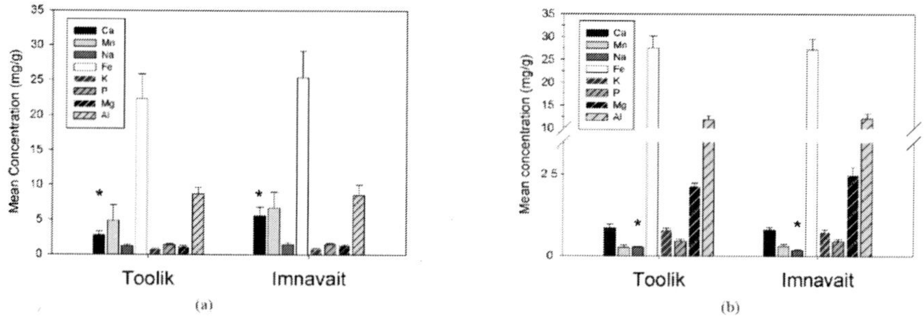

Figure 3. Differences between Toolik Lake and Imnavait Creek grids in terms of mean total concentration (mg/g) of Ca, Mn, Na, Fe, K, P, Mg and Al in (a) organic horizons and (b) mineral horizons. Asterisks indicate significant differences ($p < 0.05$).

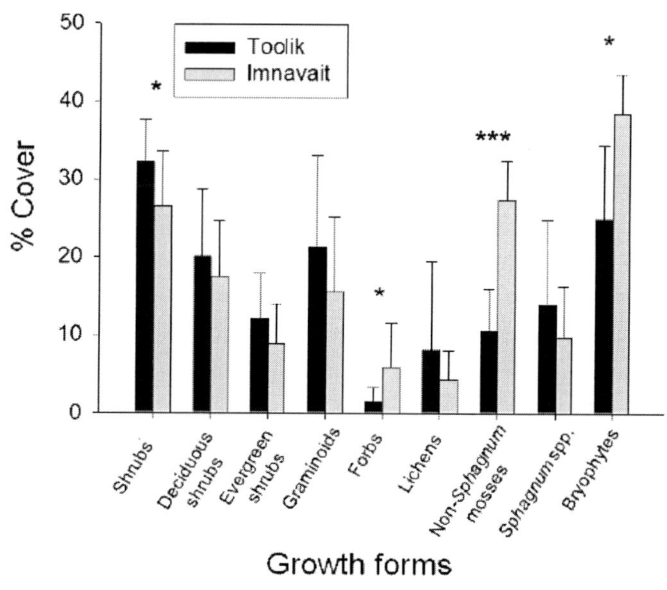

Figure 4. Differences between Toolik Lake and Imnavait Creek grids in terms of mean abundance of growth forms (% Cover). Abundance data correspond to years 2007-2008. Asterisks indicate significant differences ($p < 0.05$).

Table 2. Abundances (% Cover) of main growth forms in 24 one meter squared plots with vegetation classified as Sphagno-Eriophore- tum vaginati plant community. Abundance data correspond to years 2007-2008. Between-site comparisons were assessed with Independent Samples T-Test. Sample size within parentheses unless otherwise indicated with superscripts. SD = Standard deviation. Statistical significance achieved at p < 0.05.

Type	Growth form	Toolik		Imnavait		p-values
		Mean (n = 12)	SD	Mean (n = 12)	SD	
Vascular	Shrubs	32.25	5.36	26.51	7.04	<0.05
	Deciduous shrubs	20.12	8.61	17.56	7.21	ns
	Evergreen shrubs	12.13	5.77	8.94	5.06	ns
	Graminoids	21.31	11.77	15.63	9.52	ns
	Forbs	1.60	1.69	5.99	5.52	<0.05
Non-vascular	Lichens	8.14	11.39	4.44	3.59	ns
	Non-*Sphagnum* mosses	10.60	5.31	27.42	4.92	<0.001
	Sphagnum spp.	13.96	10.93	9.68	6.65	ns
	Bryophytes	24.99	9.45	38.50	5.07	<0.05

4. DISCUSSION

Determinations of several physico-chemical parameters, including total C, N and S and other elemental concentrations of soils occurring in Sphagno-Eriophoretum vaginati plant communities at Toolik Lake and Imnavait Creek grids, indicate that for most elements evaluated, and in spite of a 70 k year difference in landscape age, there are no significant differences between sites. In general, main patterns observed for many of the variables measured are not distant to what has been reported previously for this area [3] [5] - [7] [9] [23] [42] - [45] . Our observations confirm that both sites have similar soil acidity [3] [5] [43] and mineral soils bulk density values comparable to those reported in other studies [7] [42] . As expected, and in accordance with most of these studies, total C for both Toolik and Imnavait was 7 to 10 times larger in organic soils vs. mineral soils. Total N exhibited a similar pattern. There were no significant differences between Toolik and Imnavait in terms of the total concentration of both C and N in organic and mineral soils, hence confirming observations made by Whittinghill and Hobbie [5] . No between-site differences were found in terms of the amount of S in both organic and mineral soils, still, the concentration of this element in these soils is almost negligible.

Although some of our findings serve mostly as a confirmation of previously reported similarities in elemental concentrations between these two areas [5] [6] [24] , others represent intriguing aspects of the soil chemistry of this region that are worth highlighting. For instance, similar to findings of Keller et al. [23] , Fe had the largest concentration in both mineral and organic soils at both sites. Studies have demonstrated that Fe (as well as Al, the second highest) tends to accumulate in soils from tundra and other ecosystems [46] - [49] . Moreover, Fe is more soluble in acidic soils; therefore its availability tends to increase under these conditions [46] [50] . High concentrations of Al and Fe have been linked to substantial increases in the sorption capacity of organic soils. This situation may lead to reductions in P availability, causing plants to become P limited [47] [51] [52] . These observations suggest that due to its effect on P availability, changes in the concentration of Fe or Al in these soils could affect significantly the vegetation in this region. This is particularly relevant for wet sedge and some MAT areas where P availability is known to limit primary productivity [53] [54] .

We consider atypical the higher concentration of Ca in organic soils from Imnavait. Organic soils from both sites have similar pH. Because soil pH is known to affect the availability of Ca [12] [46] , contrasting Ca concentrations are expected to be more commonly be found between soils with significantly different acidity, as is the case for MAT vs. MNT tundra [3] [7] , and less frequently between areas with the same type of vegetation (MAT), such as Toolik and Imnavait. Accordingly, Whittinghill and Hobbie [5] found that exchangeable Ca concentrations were not significantly different between soils derived from the landscape ages represented by these sites. It is likely that higher total Ca in organic soils from Imnavait is linked to differences in landscape age between these sites. This observation would agree with findings from Keller et al. [23] which found that organic and mineral horizons from Sagavanirktok glacial surfaces had higher Ca concentration than soils from Itkillik I surfaces (not statistically corroborated). Perhaps other processes that were not evaluated in this study, such as those related to biological and/or chemical immobilization, are variable in soils from these two landscape ages, and may account for some of these differences.

However, as evidenced in this study, it is likely that the total Ca in organic soils is being influenced by between-site differences in vegetation composition, specifically, the abundance of non-Sphagnum mosses. In general, bryophytes tend to be in direct contact with underlying organic soils in these ecosystems and usually exert an influence on a number of soil processes [55] . Compared to Toolik, Imnavait had a significantly higher abundance of non-Sphagnum mosses and significantly higher concentration of Ca in organic soils, which agrees with the positive correlation found between these two parameters. Considering that both sites are influenced by similar climate and that plots are located in terrain of similar relief, dominated by MAT tundra; it seems plausible that besides landscape age, subtle variation in the abundance of non-Sphagnum mosses is contributing towards the differentiation of Sphagno-Eriophoretum vaginati plant community areas in terms of total Ca in organic soils. The work of van der Welle [56] partly support this hypothesis and revealed a significant positive correlation between non-Sphagnum moss cover and soil Ca content in several tussock tundra sites. Nevertheless, it remains to be tested if the effect of contrasting abundances on organic soil Ca concentration occurs via specific plant traits, like between-species variation in tissue concentration of this element, or via other plant community factors, like differences in species richness of non-Sphagnum mosses.

In terms of its ecosystemic effects, a higher total Ca in Imnavait soils imply that there may be less available substrate for microbial activity at that site. This is expected to diminish microbial respiration due to stabilization of organic matter by cation bridging with Ca ions [6] . However, rates of microbial activity are higher on older, more acidic landscapes [5] [8] [9] which may partly compensate for the effects of higher Ca on microbial respiration in these soils.

A significantly higher concentration of Na in minerals soils of Toolik was an unexpected finding considering that total Na have been found to be higher in mineral soils from Sagavarnirktok vs. Itkillik I glacial surfaces [23] (not statistically corroborated). However, because the dynamics of Na in mineral soils has not been characterized thoroughly in this region, we believe that between-site differences in the chemical composition of the parent material, perhaps the presence of Na-rich bedrock in sampled plots at Toolik, would be the most logical explanation for these observations. Although less likely, part of this dissimilarity might also be tied to differences in vegetation composition. Contrary to patterns observed between the abundance of non-Sphagnum mosses and total Ca in organic soils, significantly lower abundances of forbs, non-Sphagnum mosses and bryophytes in Toolik are responsible for the observed negative correlation between these growth forms and total Na in mineral soils. Assuming that forbs and bryophytes have low Na tissue concentration compared to other growth forms, it could be argued that in areas where they flourish (like Toolik), soils could reflect low Na concentration. However, in first instance, this assumption is undermined by the general scarcity of information regarding the concentration of Na in tissues of tundra plant species. Additionally, the effect of these growth forms on deeper mineral soil biogeochemical processes is expected to be negligible considering that their vertical distribution is limited to upper soil layers. Lastly, higher Na in mineral soils at Toolik might also be linked to other factors not evaluated in this study, for example, between-site variation in rates of chemical weathering of parent materials.

Soil Na has not been properly characterized in studies of soil chemistry in tundra ecosystems, except for few notable studies [3] [7] [23] . As an element, Na is present in small quantities in soils and is not an essential element for plant growth. Nevertheless, Na concentration in soils or water has been shown to promote maximal biomass yield and perform other important functional roles in plants [57] . High levels of Na could be detrimental to soil structure, soil permeability and plant growth, and in spite of its overall low levels, projected climate warming in tundra ecosystems may result in reduced soil moisture conditions [58] which may in turn affect Na dynamics in soils. Under this scenario, it becomes important to increase efforts towards a better understanding of Na dynamics in tundra soils as they are likely to change under a warming climate.

Based on our results and those of others [5] [6] [23] [42] -[44] , it is evident that landscape age differences between our sites do not result in a robust chemical imprint that might help discriminate these sites in terms of their soil physico-chemical properties. While we were able to detect significant differences in two (Ca in organic soils and Na in mineral soils) of the 11 chemical parameters evaluated, these differences are subtle compared to the conspicuous differences in concentration of other elements, like Ca, between MAT and MNT [5] [7] . It has been recognized that variation in elemental concentrations in tundra soils of the Toolik-Imnavait region are tied to larger landscape-scale factors like changes in vegetation physiognomy and pH which show considerable variation at different spatial and temporal scales in this region [1] [3] [5] [7] [9] [18] [23] . Additionally, other processes like cryoturbation and frost boil formation could cause the continuous movement and mixing of soil material [3] [15] [59] , thus preventing long term soil stability and causing continuous homogenization of nutrients throughout soil horizons [42] [60] . Freeze-thaw processes like cryoturbation are more frequent on MNT [3] [61] , though discontinuous surface organic horizons resulting from cryoturbation are common throughout the southern foothills [62] . Therefore, we suggest that while possibly occurring in a lower frequency and magnitude, these processes are at least partly responsible for the little differentiation that organic and mineral soils of these two sites reflect in terms of most of the elemental concentrations and other physico-chemical properties evaluated in this study.

Most importantly, after detecting the differences between Toolik and Imnavait in terms of the concentration of Ca in organic soils and Na in mineral soils, we were able to discern that abundance patterns of several growth forms at these sites were concordant with correlations found between these growth forms and the concentration of these elements in organic and mineral soils respectively. These observations are relevant as they demonstrate that minimal differences in soil chemistry, possibly linked to contrasting landscape ages between these two sites, are influencing aboveground patterns in vegetation composition. In turn, by directly influencing the biophysical environment through variation in moisture or temperature regimes, through differences in root mycorrhizal processes [33] , or by indirectly influencing the rates of litter input and decomposability [8] , this variation in vegetation composition could be reciprocally influencing soil chemistry, therefore perpetuating the differentiation between these two sites in terms of these two elements.

5. CONCLUSION

Investigations analyzing the relationship between soil chemistry and vegetation in tundra ecosystems in this region have shown that conspicuous soil chemical differences (e.g. pH and Ca concentration) between some different aged landscapes (e.g. Itkillik II vs. Itkillik I) are accompanied by notable differences in vegetation types (MAT vs. MNT) [3] [7] [9] . In this respect, our study indicates that soil evolution may be governed by vegetation cover as much as by time. Landscapes of different age (Itkillik I vs. Sagavariktok) and very similar vegetation type (i.e., MAT) are not distinguishable in terms of most of their soil physical and chemical properties. However, total concentration of elements like Ca and Na may vary between them. This variation, although likely linked to differences in the chemical composition of the geologic substrates underlying these soils, appears to reflect some level of agreement with abundance patterns of specific growth forms. These observations are relevant as they testify in favor of the reciprocal relationship between soil chemistry and vegetation composition in moist acidic tundra landscapes in Arctic Alaska.

ACKNOWLEDGEMENTS

We thank Amy Breen Carroll and Sayuri Ito for their work during the collection of soil samples. We are profoundly grateful to the personnel of the IITF Chemistry Laboratory for helping in the analysis and interpretation of soil chemical results. Thanks to the staff of the Toolik Lake Field Station (UAF) for their hospitality and for coordination of laboratory space. Ariel E. Lugo reviewed the original version of the manuscript and provided helpful comments. This work is based on support by the United States National Science Foundation awards OPP-0632277 and OPP-0856710. This research was conducted in cooperation with the University of Puerto Rico.

REFERENCES

1. Marion, G.M., Hastings, S.J., Oberbauer, S.F. and Oechel, W.C. (1989) Soil-Plant Element Relationships in a Tundra Ecosystem. Holarctic Ecology, 12, 296-303.

2. Walker, D.A., Auerbach, N.A. and Shippert, M.M. (1995) NDVI, Biomass, and Landscape Evolution of Glaciated Terrain in Northern Alaska. Polar Record, 31, 169-178.

3. Bockheim, J.G., Walker, D.A., Everett, L.R., Nelson, F.E. and Shiklomanov, N.I. (1998) Soils and Cryoturbation in Moist Non-Acidic and Acidic Tundra in the Kuparuk River Basin, Arctic Alaska, USA. Arctic and Alpine Research, 30, 166-174.

4. Walker, D.A. (2000) Hierarchical Subdivision of Arctic Tundra based on Vegetation Response to Climate, Parent Material and Topography. Global Change Biology, 6, 19-34.

5. Whittinghill, K.A. and Hobbie, S.E. (2011) Effects of Landscape Age on Soil Organic Matter processing in Northern Alaska. Soil Science Society of America Journal, 75, 907-917.

6. Whittinghill, K.A. and Hobbie, S.E. (2011) Effects of pH and Calcium on Soil Organic Matter Dynamics in Alaskan Tundra. Biogeochemistry, 111, 569-581.

7. Hobbie, S.E. and Gough, L. (2002) Foliar and Soil Nutrients in Tundra on Glacial Landscapes of contrasting Ages in Northern Alaska. Oecologia, 131, 453-462.

8. Hobbie, S.E. and Gough, L. (2004) Litter Decomposition in Moist Acidic and Non-Acidic Tundra with different Glacial Histories. Oecologia, 140, 113-124.

9. Hobbie, S.E., Miley, T.A. and Weiss, M.S. (2002) Carbon and Nitrogen Cycling in Soils from Acidic and Non Acidic Tundra with Different Glacial Histories in Northern Alaska. Ecosystems, 5, 761-774.

10. Walker, D.A. and Walker, M.D. (1996) Terrain and Vegetation of the Imnavait Creek Watershed. In: Reynolds, J.F. and Tenhunen, J.D., Eds., Landscape Function and Disturbance in Arctic Tundra Ecological Studies, Vol. 120, Springer-Verlag, Berlin, 73-108.

11. Walker, D.A., Epstein, H.E., Jia, G.J., Balser, A., Copass, C., Edwards, E.J., Gould, W.A., Hollingsworth, J., Knudson, J., Maier, H.A., Moody, A. and Raynolds, M.K. (2003) Phytomass, LAI, and NDVI in Northern Alaska: Relationships to Summer Warmth, Soil pH, Plant Functional Types, and Extrapolation to the Circumpolar Arctic. Journal of Geophysical Research, 108, 8169-8185.

12. Gough, L., Shaver, G.R., Carroll, J., Royer, D.L. and Laundre, J.A. (2000) Vascular Plant Species Richness in Alaskan Arctic Tundra: The Importance of Soil pH. Journal of Ecology, 88, 54-66.

13. Eskelinen, A., Stark, S. and Männistö, M. (2009) Links between Plant Community Composition, Soil Organic Matter Quality and Microbial Communities in Contrasting Tundra Habitats. Oecologia, 161, 113-123.

14. Hobbie, S.E. (1992) Effects of Plant Species on Nutrient Cycling. Trends in Ecology & Evolution, 7, 336-339.

15. Walker, D.A., Epstein, H.E., Gould, W.A., Kelley, A.M., Kade, A.N., Knudson, J.A., Krantz, W.B., Michaelson, G., Peterson, R.A., Ping, C.L., Raynolds, M.K., Romanovsky, V.E. and Shur, Y. (2004) Frost-Boil Ecosystems: Complex Interactions between Landforms, Soils, Vegetation and Climate. Permafrost and Periglacial Processes, 15, 171-188.

16. Hobbie, S.E. (1996) Temperature and Plant Species Control over Litter Decomposition in Alaskan Tundra. Ecological Monographs, 66, 503-522.

17. Chu, H. and Grogan, P. (2010) Soil Microbial Biomass, Nutrient Availability and Nitrogen Mineralization Potential among Vegetation-Types in a Low Arctic Tundra Landscape. Plant and Soil, 329, 411-420.

18. Walker, M.D., Walker, D.A. and Auerbach, N.A. (1994) Plant Communities of a Tussock Tundra Landscape in the Brooks Range Foothills, Alaska. Journal of Vegetation Science, 5, 843-866.

19. Walker, D.A., Walker, M.D., Gould, W.A., Mercado, J., Auerbach, N.A., Maier, H.A. and Neufeld, G.P. (2010) Maps for Monitoring Long-Term Changes to Vegetation Structure and Composition, Toolik Lake, Alaska. In: Bryn, A., Dramstad, W. and Fjellstad, W., Eds., Mapping and Monitoring of Nordic Vegetation and Landscapes, Vol. 1, Norsk Institutt for Skog og Landskap, Ås, Norway, 121-123.

20. Walker, D.A., Lederer, N.D. and Walker, M.D. (1987) Permanent Vegetation Plots: Site Factors, Soil Physical and Chemical Properties and Plant Species Cover. Department of Energy, R4D Program Data Report, Plant Ecology Laboratory, Institute of Arctic and Alpine Research, Boulder, National Snow and Ice Data Center. Identifier Number: ARCSS110.

21. Walker, D.A. and Barry, N. (1991) Toolik Lake Permanent Vegetation Plots: Site Factors, Soil Physical and Chemical Properties, Plant Species Cover, Photographs, and Soil Descriptions. Data Report 48, Department of Energy R4D Program, Institute of Arctic and Alpine Research, University of Colorado, Boulder.

22. Walker, D.A., Binnian, E., Evans, B.M., Lederer, N.D., Nordstrand, E. and Webber, P.J. (1989) Terrain, Vegetation and Landscape Evolution of the R4D Research Site, Brooks Range Foothills, Alaska. Holarctic Ecology, 12, 238-261.

23. Keller, K., Blum, J.D. and Kling, G.W. (2007) Geochemistry of Soils and Streams on Surfaces of Varying Ages in Arctic Alaska. Arctic, Antarctic, and Alpine Research, 39, 84-98.

24. Walker, D.A. and Maier, H.A. (2008) Vegetation in the Vicinity of the Toolik Field Station, Alaska. Biological Papers of the University of Alaska 28, Institute of Arctic Biology, Fairbanks.

25. Wahrhaftig, C. (1965) Physiographic Divisions of Alaska U.S. Geological Survey Professional Paper 482. US Government Printing Office, Washington DC.

26. Osterkamp, T.E., Petersen, J.K. and Collet, T.S. (1985) Permafrost Thicknesses in the Oliktok Point, Prudhoe Bay and Mikkelsen Bay Areas of Alaska. Cold Regions Science and Technology, 11, 99-105.

27. Hamilton, T.D. (2003) Glacial Geology of Toolik Lake and the Upper Kuparuk River Region. Biological Papers of the University of Alaska No. 26, University of Alaska Printing Services, Fairbanks.

28. Hamilton, T.D. (2003) Surficial Geology of the Dalton Highway (Itkillik-Sagavanirktok Rivers) Area, Southern Arctic Foothills, Alaska. Alaska Division of Geological & Geophysical Surveys Professional Report 121, Alaska, 32 p.

29. Cherry, J., Déry, S.J., Cheng, Y., Stieglitz, M., Jacobs, M.S. and Pan, F. (2014) Climate and Hydrometeorology of the Toolik Lake Region and the Kuparuk River Basin: Past, Present and Future. In: Hobbie, J.E. and Kling, G.W., Eds., Alaska's Changing Arctic: Ecological Consequences for Tundra, Streams and Lakes, Oxford University Press, New York, 31-60.

30. Gould, W.A. and Mercado-Díaz, J.A. (2014) Decadal-Scale Changes of Vegetation from Long-Term Plots in Alaskan Tundra. In: Hobbie, J.E. and Kling, G.W., Eds., Alaska's Changing Arctic: Ecological Consequences for Tundra, Streams and Lakes, Vignette 5.5, Oxford University Press, New York, 130-131.

31. Walker, D.A., Hamilton, T.D., Maier, H.A., Munger, C.A. and Raynolds, M.K. (2014) Glacial History and Long-Term Ecology in the Toolik Lake Region. In: Hobbie, J.E. and Kling, G.W., Eds., Alaska's Changing Arctic: Ecological Consequences for Tundra, Streams and Lakes, Oxford University Press, New York, 61-80.

32. Shaver, G.R., Laundre, J.A., Bret-Harte, M.S., Chapin III, F.S., Mercado-Díaz, J.A., Giblin, A.E., Gough, L., Gould, W.A., Hobbie, S.E., Kling, G.W., Mack, M.C., Moore, J.C., Nadelhoffer, K., Rastetter, E.B. and Schimel, J.P. (2014) Terrestrial Ecosystems at Toolik Lake, Alaska. In: Hobbie, J.E. and Kling, G.W., Eds., Alaska's Changing Arctic: Ecological Consequences for Tundra, Streams and Lakes, Oxford University Press, New York, 90-142.

33. Walker, D.A. (1996) GIS Data from the Alaska North Slope. National Snow and Ice Data Center.

34. Mercado-Díaz, J.A. (2011) Plant Community Responses of the Alaskan Arctic Tundra to Environmental and Experimental Changes in Climate. M.Sc. Thesis, University of Puerto Rico, Río Piedras Campus, San Juan.

35. LECO Corp. (2006) LECO TruSpec CN Carbon/Nitrogen Determinator Instruction Manual. St. Joseph.

36. LECO Corp. (2005) Carbon and Nitrogen in Soil and Sediment. Organic Application Note: TruSpec CN (Form No. 203-821-275). St. Joseph.

37. LECO Corp. (2006) LECO TruSpec Add-On Module Sulfur Analyzer Instruction Manual. St. Joseph.

38. LECO Corp. (2008) Sulfur in Cement, Fly Ash, Limestone, Soil and Ore. Organic Application Note: TruSpec S (Form No. 203-821-345). St. Joseph.

39. Luh Huang, C.Y. and Schulte, E.E. (1985) Digestion of Plant Tissue for Analysis by ICP Emission Spectroscopy. Communications in Soil Science and Plant Analysis, 16, 943-958.

40. Wilde, S.A., Corey, R.B., Iyer, J.G. and Voight, G.K. (1979) Soil and Plant Analysis for Tree Culture. 5th Edition, Oxford & IBH Publishing Co., New Delhi.

41. McLean, E.O. (1982) Soil pH and Lime Requirement. In: Page, A.L., Miller, R.H. and Keeney, D.R., Eds., Methods of Soil Analysis, Part 2, Chemical and Microbiological Properties, Agronomy Monograph Number 9, Soil Science Society of America, Madison, 199-224.

42. Michaelson, G.J., Ping, C.L. and Kimble, J.M. (1996) Carbon Storage and Distribution in Tundra Soils of Arctic Alaska, U.S.A. Arctic and Alpine Research, 28, 414-424.

43. Ping, C.L., Michaelson, G.J., Loya, W.M., Chandler, R.J. and Malcolm, R.L. (1997) Characteristics of Soil Organic Matter in Arctic Ecosystems of Alaska. In: Lal, R., Kimble, J.M., Follet, R.F.

and Stewart, B.A., Eds., Soil Processes and the Carbon Cycle, CRC Press LLC, Boca Raton, 157-167.

44. Ping, C.L., Bockheim, J.G., Kimble, J.M., Michaelson, G.J. and Walker, D.A. (1998) Characteristics of Cryogenic Soils along a Latitudinal Transect in Arctic Alaska. Journal of Geophysical Research, 103, 28917-28928.

45. Giblin, A.E., Nadelhoffer, K.J., Shaver, G.R., Laundre, J.A. and McKerrow, A.J. (1991) Biogeochemical Diversity along a Riverside Toposequence in Arctic Alaska. Ecological Monographs, 61, 415-435.

46. Moore, P.D. (2008) Tundra. Infobase Publishing, New York.

47. Giesler, R., Petersson, T. and Högberg, P. (2002) Phosphorus Limitation in Boreal Forests: Effects of Aluminum and Iron Accumulation in the Humus Layer. Ecosystems, 5, 300-314.

48. Birkeland, P.W., Burke, R.M. and Benedict, J.B. (1989) Pedogenic Gradients for Iron and Aluminum Accumulation and Phosphorus Depletion in Arctic and Alpine Soils as a Function of Time and Climate. Quaternary Research, 32, 193-204.

49. Ugolini, F.C., Stoner, M.G. and Marrett, D.J. (1987) Arctic Pedogenesis: 1. Evidence for Contemporary Podzolization. Soil Science, 144, 90-100.

50. Brady, N.C. and Weil, R.R. (2008) The Nature and Properties of Soils. Prentice-Hall Inc., New Jersey.

51. Giesler, R., Andersson, T., Lövgren, L. and Persson, P. (2005) Phopshate Sorption in Aluminum and Iron-Rich Humus Soils. Soil Science Society of America Journal, 69, 77-86.

52. Kang, J., Hesterberg, D. and Osmond, D.L. (2009) Soil Organic Matter Effects on Phosphorus Sorption: A Path Analysis. Soil Science Society of America Journal, 73, 360-366.

53. Shaver, G.R. and Chapin III, F.S. (1986) Effect of Fertilizer on Production and Biomass of Tussock Tundra, Alaska, U.S.A. Arctic and Alpine Research, 18, 261-268.

54. Shaver, G.R. and Chapin III, F.S. (1995) Long-Term Responses to Factorial, NPK Fertilizer Treatment by Alaskan Wet and Moist Tundra Sedge Species. Ecography, 18, 259-275.

55. Jäggerbrand, A.K., Björk, R.G., Callaghan, T. and Seppelt, R.D. (2011) Effects of Climate Change on Tundra Bryophytes. In: Tuba, Z., Slack, N.G. and Stark, L.R., Eds., Bryophyte Ecology and Climate Change, Cambridge University Press, New York, 211-236.

56. van der Welle, M.E.W., Vermeulen, P.J., Shaver, G.R. and Berendese, F. (2003) Factors Determining Plant Species Richness in Alaskan Arctic Tundra. Journal of Vegetation Science, 14, 711-720.

57. Subberao, G.V., Ito, O., Berry, W.L. and Wheeler, R.M. (2003) Sodium—A Functional Plant Nutrient. Critical Reviews in Plant Sciences, 22, 391-416.

58. Xu, W., Yuan, W., Dong, W., Xia, J., Liu, D. and Chen, Y. (2013) A Meta-Analysis of the Response of Soil Moisture to Experimental Warming. Environmental Research Letters, 8, 1-8.

59. Munroe, J.S. and Bockheim, J.G. (2001) Soil Development in Low-Arctic Tundra of the Northern Brooks Range, Alaska, U.S.A. Arctic, Antarctic and Alpine Research, 33, 78-87.

60. Michaelson, G.J., Ping, C.L. and Kimble, J.M. (2001) Effects of Soil Morphological and Physical Properties on Estimation of Carbon Storage. In: Lal, R., Kimble, J.M., Follett, R.F. and Stewart, B.A., Eds., Assessment Methods for Soil Carbon, Lewis Publishers, Boca Raton, 339-347.

61. Bockheim, J.G., Walker, D.A. and Everett, L.R. (1997) Soil Carbon Distribution in Non Acidic and Acidic Tundra of Arctic Alaska. In: Lal, R., Kimble, J.M., Follett, R.F. and Stewart, B.A., Eds., Soil Processes and the Carbon Cycle, CRC Press, Boca Raton, 143-155.

62. Ping, C.L., Clark, M.H. and Swanson, D.K. (2004) Cryosols in Alaska. In: Kimble, J.M., Ed., Cryosols, Permafrost-Affected Soils, Springer-Verlag, New York, 71-94.

CHAPTER 2

Planting Cotton in a Crop Residue in a Semiarid Climate: Water Balance and Lint Yield

Robert J. Lascano[1], Dan R. Krieg[2], Jeffrey T. Baker[3], Timothy S. Goebel[1], Dennis C. Gitz III[1]

[1]Wind Erosion and Water Conservation, USDA-ARS*, Lubbock, TX, USA

[2]Department of Plant and Soil Science, Texas Tech University, Lubbock, TX, USA

[3]Wind Erosion and Water Conservation, USDA-ARS, Big Spring, TX, USA

ABSTRACT

Cotton (Gossypium hirsutum L.) is planted on more land area than any other crop on the Texas High Plains. Much of this area is considered highly erodible and requires a conservation compliance program to participate in government farm programs. Because this region is semiarid and because irrigation water is increasingly limited, water conservation and efficient use of water are necessary to maximize cotton lint yields. One popular conservation compliance practice used is to plant cotton into a chemically terminated small grain crop, i.e., residue that provides wind protection to the cotton seedlings. Our hypothesis was that in a semiarid region the use of a small grain cover crop under irrigated conditions would use more water than it conserves compared to conventional tilled cotton, thus reducing cotton lint yields. To test the hypothesis separate field studies over two growing seasons and on two soil textures, a loamy fine sand and a clay loam, were conducted. The main treatments were tillage systems (conventional and conservation using terminated wheat residue). The two split plot treatments were water supply based on replacement of calculated grass reference evapotranspiration (ET_o). Tillage did not affect the amount of water used by the cotton crop at either location (< 7% difference, $P > 0.05$) except for an 80% ET_o irrigation treatment at a single location where the bare soil treatment used 10% more water than the residue treatments for both years. The residue treatment decreased ($P < 0.05$) cotton lint yields at both locations by 12% except for the 50% ET_o single irrigation treatment in which the residue treatment yielded 14% more lint than the bare soil treatment. The use of terminated wheat residue had no impact on soil water storage during any part of the year. During a 5-month period associated with wheat growth, the wheat evapotranspiration was 20 to 40 mm more water ($P < 0.05$) than that lost through soil water evaporation from the conventional treatments. The use of terminated wheat residue did not benefit the water balance of the cotton crop, and was associated with decreased cotton lint yields. The results were consistent with our working hypothesis, and disproved the idea that planting cotton into wheat stubble cover increases water use efficiency.

KEYWORDS

Soil Water, Evapotranspiration, Cropping Systems, Semi-Arid, Wheat Residue

1. INTRODUCTION

In 2013 and 2014, about 1.2 million ha were planted with upland cotton on the Texas South Plains [1] . The annual mean precipitation in this region is 467 ± 159 mm and about 50% of this rainfall occurs during the 4-month (May to August) growing season. About half of the planted area is irrigated with a water supply ranging from 2.5 to 7.5 $mm \cdot d^{-1}$ and the potential evaporative demand is on average 9.8 ± 0.8 $mm \cdot d^{-1}$ during the four-month growing season [2] . The source of the irrigation water is the Ogallala aquifer, where withdrawals exceed natural recharge, resulting in a declining water table [3] . Much of the water applied to the cotton is lost to bare soil water evaporation (E_{soil}) in conventional tillage systems [4] [5] . In this semiarid region, where cotton is planted in a wide row spacing (1-m) the crop seldom reaches a leaf area index of 3 $m^2 \cdot m^{-2}$ that is necessary to achieve canopy closure [6] [7] . Furthermore this exposed and bare soil is conducive to increase the amount of E_{soil} [8].

Given the decline of irrigation water from the Ogallala aquifer [3] [9] with a decreasing pumping capacity, particularly during periods of low rainfall which are frequent in this region, led to develop agronomic practices to maximize cotton's water use efficiency (WUE), i.e., production of lint yield per unit of water taken up by the plant [10] . An example of this conservation practice is to plant cotton into a winter wheat (Triticum aestivum L.) crop that is chemically terminated in the spring before the cotton crop is planted [11] . Residues may protect cotton seedlings by providing shelter to wind damage and sand abrasion [12] . Also, the residue may reduce E_{soil}; however, in some cases it may use more water than it conserves [5] [8] via wicking [13] . This is the process whereby the wheat residues may contribute to additional water loss via capillary action, i.e., wicking effect [14] . In this situation, the wheat residue may increase the ability to transport water from the seedbed to the atmosphere via evaporation and thus resulting in a reduction of soil water and negatively affecting cotton emergence when planted in the wheat residue.

In the cotton cropping systems of the Texas High Plains, the effect of crop residues on water conservation and lint yield, are a function of the water supply, from irrigation and rainfall, soil type, and tillage practices [15] [16] . Planting cotton into a wheat residue on a sandy loam soil only increased cotton lint yield when the crop was irrigated at the grass reference evapotranspiration demand (ET_o). However, this practice is not suited to the availability of irrigation water from the Ogallala aquifer. In field trialsand over a five-year period, in a loamy fine sandy soil the residue did not impact lint yield under irrigated conditions, but significantly and adversely affected dryland lint yields. In a clay loam soil, the lint yield of cotton planted in a wheat residue increased on average by 13% under irrigation and reduced dryland cotton lint yield by 8% [17] [18] . From these results, it can be concluded that the effects of wheat residues on cotton lint yield depend primarily on the water supply. The texture effect is related to the amount of water that can be stored in the soil, particularly from the time the wheat is terminated to when the cotton is planted. For example, in the surface 0.20 m of the soil profile, a clay soil can store about twice as much water as a sandy soil and for dryland conditions the additional stored water would be available during cotton planting and favor seed germination and emergence of the cotton plant.

In first analysis for the Texas High Plains, the amount of water required to establish the wheat residue is often ignored in the water balance of dryland and limited irrigated cotton cropping systems [4] [5] [8] . To establish a wheat residue that would not impact the germination and emergence of the subsequent cotton crop planted in the residue, and of sufficient growth to be beneficial to cotton seedlings requires a net gain of water. In the Texas High Plains, winter wheat is planted in months (November to February) where the average monthly rainfall is 16 ± 2 mm [2] . Clearly, in a dryland system the establishment of the residue comes with the risk of using stored water that otherwise would be available at the time needed when the cotton crop is planted [19] . Furthermore, the winter wheat crop requires water to grow and be of benefit to the cotton crop. The measured daily ET rate of irrigated winter wheat varied between 1 and 2 $mm \cdot d^{-1}$ with a maximum of 13 $mm \cdot d^{-1}$ during the late spring [20] . Therefore, it was concluded that for the Texas High Plains under dryland conditions the establishment of a wheat residue is perhaps not practical and will require irrigation to be of value [19] .

The short- and long-term benefits of residues on soil physical and chemical properties, and crop production has been widely investigated, e.g., [15] [21] - [23] . Residues increase infiltration of water, from either rain or irrigation, by increasing the surface roughness and flow path tortuosity and thus increasing the time for more water to infiltrate by surface ponding [19] [24] [25] . Residues, also modify the kinetic energy from raindrops impacting the soil surface [26] [27] , which leads to less erosion [26] and increasing rain interception [28] by as much as 10% of the annual rainfall [29] . A long-term benefit of residues is to increase water storage and thus providing more water for crop production [30] [31] .

The impact of residues on E_{soil} is well documented [4] [5] [8] [32] - [34] . Residues can reduce the rate of E_{soil} in the so-called first stage [25] [30] [34] - [37] . The type of residue also affects the E_{soil} rate and wheat, for example, is more effective than cotton [35] . The presence of a residue reduces E_{soil}; however, the total evaporation equilibrates between residue covered soil and bare soil after a long period of time between water inputs [35] . For example, the rate of E_{soil} from a bare soil and a residue-covered soil will be different; however, the total amount of water lost from the bare and residue-covered soil will be very similar after a long period of time [5] [8] . The rate of E_{soil} is defined by exchanges of energy and water between the soil surface and the surrounding environment [32] and determined by the combined water and energy balance of the soil surface. The aerodynamic resistance to vapor and latent heat flux from the soil surface to the atmosphere is increased by the presence of a residue, which decreases the rate of E_{soil} [5] [8] [38] [39] . Another effect of a residue is to reduce the amount of irradiance that reaches the surface, decreasing soil temperature and E_{soil}.

The presence of a surface residue modifies the E_{soil} and affects the WUE [8] . The daily E_{soil} and crop transpiration (T) for a 100-day growing season of cotton planted into terminated wheat and conventionally tilled soil where compared by Lascano et al. [8] . In this experiment daily E_{soil} was measured for a seven-day period using microlysimeters [32] and calculated using the mechanistic ENWATBAL model [4] [40] . Both cotton cropping systems, residue and conventional, had the same seasonal ET. However, the conventional tilled cotton evaporated more E_{soil} and had a lower amount of plant T than the cotton planted into the terminated wheat. The wheat residue increased the seasonal crop T, i.e., more water used for plant growth, and reduced the seasonal E_{soil}, resulting in an increase of lint yield of 35% compared to the cotton planted in the conventional tillage with the same amount of water. The WUE of the cotton planted in the wheat residue was 2.6 g lint per kg of water used in ET, which was 27% more lint than the conventionally planted cotton [8] . In these field experiments the amount of water used to grow and establish the wheat residue were not measured and thus the calculated value of WUE is biased towards the cotton lint yield from the residue.

Our hypothesis is that despite the many benefits that residue offer, in a semiarid environment establishing the residue would use more water than conserved when used along a cotton crop. In our hypothesis the assumption was made that the cover crop that provides the residue and cotton were supplemented as needed with irrigation and each tillage treatment received the same amount of water annually. Further, it was postulated that the use of chemically terminated winter wheat residue would not significantly increase lint yields because less water would available for cotton growth and lint production. The main objective of this study was to examine and compare the water balance of two cotton cropping systems for a two-year period for conventionally tilled cotton, and for cotton planted into terminated wheat. Specifically, our objectives were to 1) compare the water used by wheat vs. soil water lost to evaporation from the conventionally tilled bare soil; 2) determine if wicking occurred after the wheat was terminated and before the cotton was planted; and 3) determine the effects that bare soils and residue-covered soils have on the water available for cotton growth and lint yield.

2. MATERIALS AND METHODS

This study was done during two growing seasons, 1994 and 1995, at experimental fields within two research farms managed by Texas Tech University, Lubbock, TX. One farm was in central Terry County, TX near Brown- field (33°10'47"N - 102°16'15"W) where the site has an Amarillo loamy fine sand soil [fine-loamy, mixed, thermic, Torrertic Paleustalf]. The other farm was in northern Lubbock County, TX near

New Deal (33°44'13.76"N, 101°43'58.04"W) with a Pullman clay loam soil [fine, mixed, thermic, Torrertic Paleustoll]. These two research farms were part of a long-term (1991-2001) study to evaluate dryland cropping systems, cotton and sorghum (Sorghum bicolor L.) in the semiarid climate of the Texas High Plains. Results presented here are partially based on work by Vorheis [41] and Ralston [42].

2.1. Irrigation

Irrigation-water at each location was applied using sprinkler irrigation methods. At Brownfield, TX a center pivot, 400 m long, irrigation system provided 50, 75, and 100% of the weekly crop water use with Low Energy Precision Application (LEPA) spray nozzles [43] . The linear (300 m in length) irrigation system used at New Deal provided 40% or 80% of weekly crop water use with LEPA spray nozzles. Half the field at New Deal was irrigated at 40% ET_o and the other half irrigated at 80% ET_o.

At each location, the amount of water applied was based on calculated daily grass reference ET_ousing as weather input net irradiance, air temperature and humidity, and wind speed measured at each site, and using the Penman-Monteith method as given by Allen et al. [44] . The daily water use was determined as ET_o multiplied by a locally developed crop coefficient (K_c) that depended on the ground cover [45] . The weekly crop water use was calculated using the appropriate ET_o and $K_c$for that site. Date and amount of water applications for Brownfield are given in Table 1 and for New Deal in Table 2. Rainfall was measured at each site using a tipping bucket. Irrigation treatments at both sites did not begin until cotton emergence. The irrigation levels were different between the two locations because the two studies were independent of each other, with different soil textures and environmental conditions.

2.2. Experimental Design

The experimental design at both locations was a randomized block, split plot setup. The experimental field at each site was about 46 ha. The major plots were tillage effects and the irrigation treatments were the split plots. All treatments were replicated four times and the plot size of each replicate was 156 ´ 183 m. The irrigation treatments (i.e., 50% and 100% ET_o at Brownfield, and 40% and 80% ET_o at New Deal) were started after the cotton emerged (late May through early June). Therefore, all plots for each location received identical irrigation amounts between cotton harvest from the previous season until cotton emergence the next season. One half of each irrigation treatment at each location was planted conventionally with cotton (left fallow in winter) and the other half was planted with cotton in terminated winter wheat. Locations were previously used for cotton production prior to this study with half of each field in terminated wheat residue and half of each field under conventional tillage. Treatments were replicated, n = 4, and mean separation was done using the PROC MIXED procedure and with t-type confidence interval of 0.05 using SAS on an Apple^{Ò} computer (WMware and SAS version 9.2).

2.3. Cultural Practices

Winter wheat at both locations was drilled at a rate of 30 kg·ha^{-1} at 0.2-m row spacing into listed soil resulting in two rows of wheat in each furrow on 0.75-m centers. The tops of the beds were left bare in preparation for the planting of cotton. The wheat (TAM-200, Texas A&M University Foundation Seed, College Station, TX)[#] grew until it was tall, about 0.3 m, enough to provide wind protection, which was usually in mid- to late-April and was then terminated with the herbicide glysophate (Roundup^{Ò}, Monsanto Company, St. Louis, MO) at a rate of 2 L·ha^{-1}. Cotton (HS-26, Paymaster Technology Corp., Scott, MS) was planted into the terminated wheat and on the fallow plots during early to mid-May. A complete fertilizer blend consisting of 50-50-0-10 was applied to the entire area after each cotton harvest and prior to planting the wheat in the fall each year. Supplemental N was provided through the irrigation water at a rate of 0.2 kg N mm^{-1}·ha^{-1} with an irrigation volume of 25 mm for the 100%

and 80% irrigation treatments through the third week of flowering. Important dates for different cultural operations at both locations are given in Table 3.

2.4. Measurements

Soil water contents were measured gravimetrically at various times throughout the year, usually one to two times per month [46] , and these were converted to volumetric water content using appropriate bulk density values provided by Baumhardt et al. [47] . On each sampling date, 6 - 10 soil samples per plot were taken in 0.30-m increments to a 1.2-m depth. Water used through E_{soil} and/or T during different crop stages was based on the water balance, inputs and outputs and included precipitation, irrigation and net change in soil water content. When no crops were on the ground, all the water loss was assumed to be through E_{soil}, and if a crop did exist, the water loss was assumed to be a combination of E_{soil} and T [48] . Specifically, the following quantities were determined: 1) water required to grow an adequate amount of residue; 2) water used by the wheat after being chemically terminated; 3) water lost through E_{soil}; 4) water used by each cotton crop; and 5) the effect of each cotton cropping system on cotton lint yields. Runoff was assumed to be negligible because of slopes $< 0.1\%$ and soil water content below the 1.2-m depth was not measured because of the existence of a caliche layer around the 1.4-m depth. Cotton lint yield was estimated by hand harvesting 10 m^2 of plot area in each replication of each treatment. Cotton yield components were determined and their relative contribution to total lint yield evaluated.

2.5. Additional Experiment

An additional experiment was done in New Deal, TX in 1996 to measure evaporative losses of cotton planted in a bare soil and wheat terminated residue, without any irrigation. To measure the loss of water by evaporation, gravimetric soil samples were taken in the top 0.60 m of the soil profile in about five-day increments starting on the day the wheat was terminated by applying glysophate, 15 May 1996, and ending on the 3 June 1996, when more than 75% of the cotton planted had emerged. Gravimetric soil water content was converted to a volume basis using the measured soil bulk density and these measurements were replicated six times. The bare soil and wheat residue treatment was replicated three times and each plot was the same size as previously used, i.e., 156 ′ 183 m.

Table 1. Irrigation applied and rain for two irrigation treatments and two growing seasons in Brownfield, TX.

Date	Irrigation Treatment [mm]		Rain [mm]	Date	Irrigation Treatment [mm]		Rain [mm]
	50% ET_o	100% ET_o			50% ET_o	100% ET_o	
28-Apr-94			7.4	04-Jan-95			4.6
29-Apr-94	25.4	25.4		20-Jan-95			20.1
29-Apr-94			8.4	27-Jan-95			3.3
30-Apr-94			18.0	18-Feb-95			13.0
3-May-94			1.0	01-Mar-95			1.0
10-May-94			15.5	10-Mar-95			5.6
11-May-94			29.2	07-Apr-95			8.1
12-May-94			54.1	09-Apr-95	25.4	25.4	
13-May-94			32.3	15-Apr-95	25.4	25.4	
18-May-94	6.4	6.4		02-May-95			1.3
25-May-94			3.3	05-May-95			10.9
26-May-94			1.5	06-May-95			4.1
27-May-94			5.6	11-May-95			11.4
31-May-94	25.4	25.4		15-May-95			1.5
3-Jun-94	13.2	13.2		24-May-95			1.1
10-Jun-94	25.4	25.4		30-May-95			41.0
12-Jun-94			1.5	04-Jun-95			6.4
21-Jun-94	25.4	25.4		10-Jun-95			13.5
12-Jul-94	10.8	21.6		23-Jun-95			18.0
13-Jul-94			1.8	24-Jun-95			7.0
14-Jul-94			3.8	27-Jun-95			21.0
22-Jul-94	25.4	50.8		05-Jul-95	14.0	28.0	
28-Jul-94			14.5	15-Jul-95			1.3
30-Jul-94			14.5	18-Jul-95	10.0	20.3	
3-Aug-94			3.0	20-Jul-95			5.3
8-Aug-94	25.4	50.8		21-Jul-95			2.8
27-Aug-94	12.5	25.4		25-Jul-95	10.0	20.3	
1-Sep-94			11.7	28-Jul-95	6.3	12.7	
9-Sep-94			2.0	31-Jul-95			1.3
15-Sep-94			3.8	01-Aug-95			5.1
16-Sep-94			7.9	14-Aug-95			58.0
7-Oct-94			10.7	21-Aug-95			6.0
8-Oct-94			5.3	09-Sep-95			19.3
15-Oct-94			3.6	10-Sep-95			13.7
18-Oct-94			2.3	13-Sep-95			3.0
5-Nov-94			5.8	16-Sep-95			134.0
24-Nov-94			4.3	22-Sep-95			11.0
				28-Sep-95			5.0
Total [mm]	**195.3**	**269.8**	**272.8**		**91.1**	**132.1**	**458.7**

Table 2. Irrigation applied and rain for two irrigation treatments and two growing seasons, in New Deal, TX.

Date	Irrigation Treatment [mm]		Rain [mm]	Date	Irrigation Treatment [mm]		Rain [mm]
	40% ET_o	80% ET_o			40% ET_o	80% ET_o	
26-May-94			14.0	22-Jan-95			6.9
13-Jun-94	25.4	25.4		29-Jan-95			3.0
16-Jun-94			3.0	10-Feb-95	38.1	38.1	
29-Jun-94			19.0	18-Feb-95			3.3
13-Jul-94			51.0	5-Mar-95			1.5
14-Jul-94			10.0	12-Mar-95			6.9
19-Jul-94	25.4	50.8		18-Mar-95			1.3
21-Jul-94			5.0	25-Mar-95			1.3
2-Aug-94			1.0	1-Apr-95			2.3
11-Aug-94	30.8	63.5		8-Apr-95			18.3
14-Aug-94			3.6	15-Apr-95	50.8	50.8	
8-Sep-94			7.4	16-Apr-95			8.1
14-Sep-94			11.4	18-Apr-95			1.0
15-Sep-94			8.4	5-May-95			18.5
7-Oct-94			4.3	6-May-95			5.1
14-Oct-94			1.5	15-May-95			14.5
15-Oct-94			4.6	23-May-95			2.3
18-Oct-94			1.3	24-May-95			1.8
19-Oct-94			3.3	25-May-95			2.0
20-Oct-94			1.8	26-May-95			5.1
4-Nov-94			12.7	29-May-95			16.3
19-Nov-94			2.0	30-May-95			7.9
6-Dec-94			1.0	2-Jun-95			1.0
27-Dec-94			1.3	3-Jun-95			6.4
30-Dec-94			1.0	10-Jun-95			13.5
				2-Jul-95			14.5
				14-Jul-95	16.5	33.0	
				18-Jul-95			3.3
				24-Jul-95	10.1	20.3	
				26-Jul-95	10.1	20.3	
				31-Jul-95			4.8
				1-Aug-95			2.7
				2-Aug-95			1.3
				19-Aug-95	25.4	50.8	
				9-Sep-95			2.8
				10-Sep-95			13.2
				12-Sep-95			13.5
				15-Sep-95			90.0
				19-Sep-95			55.0
				29-Sep-95			100.0
Total [mm]	**81.6**	**139.7**	**168.6**		**151.0**	**213.3**	**449.4**

Table 3. Cultural operations at two sites and growing seasons.

Cultural Practice	New Deal	Brownfield
Wheat Planted	3-Dec-93	15-Dec-93
	1-Dec-94	4-Dec-94
Wheat Terminated	4-May-94	28-Apr-94
	26-Apr-95	21-Apr-95
Cotton Planted	20-May-94	5-May-94
	15-May-95	12-May-95
Cotton Harvested	6-Nov-94	7-Nov-94
	10-Nov-95	2-Nov-94

3. RESULTS AND DISCUSSION

3.1. Wheat Water Use

The first research objective was to measure and compare the water used by the wheat to the soil water lost to evaporation from conventionally tilled bare soil. The amount of water used to establish a wheat residue and the amount of water lost to bare E_{soil} for Brownfield and New Deal are given inTable 4. During the wheat growth stage, all plots received the same irrigation amount. These results showed that in 1995, the wheat used more ($P \leq 0.05$) water than the bare soil lost E_{soil} in both treatments and locations. At Brownfield, the wheat used 40% more water ($P \leq 0.05$) than the bare soil lost to water evaporation in the loamy fine sand textured soil. At New Deal, the wheat used 36% more water ($P \leq 0.05$) than the bare soil lost to evaporation in the clay loam soil. During the spring of 1995, establishing a wheat residue required more water compared to leaving the soil bare at both locations and on both soil textures.

The water depth in the upper 0.6 m of the soil profile at wheat planting and at wheat termination at both sites, Brownfield and New Deal, TX are given in Table 5. Even though all the plots within each location received the same amount of irrigation during the wheat growth period, our measurements demonstrated the variability in soil water content at each location. The coefficient of variation of the measured mean soil water content was > 25% (data not shown). This variability could be explained by the irrigation treatments applied during the previous season because overall soil water content could increase over the growing season in the highest irrigation treatments compared to the lower ones.

Table 4. Water balance (inputs and outputs) at two locations for the 1995-growing season.

Year	Brownfield				New Deal			
	Inputs [mm]		Outputs* [mm]		Inputs [mm]		Outputs* [mm]	
			Residue	Treatment			Residue	Treatment
	Irrigation*	Rain	Bare Soil	Residue	Irrigation*	Rain	Bare Soil	Residue
1995	53.0	62.0	94.5 (17)	132.5 (16)	89.0	55.0	105 (12)	142.5 (13)

*Irrigation treatments were not active during the wheat growth; *Calculated values of outputs are the mean of all measurements and the standard deviations are given in parenthesis.

At both locations, the bare soil treatment resulted in an increase of the soil water content in the top 0.6 m by > 30% ($P \leq 0.05$) between the time when the wheat was terminated and planted (Table 5). However, the wheat residue treatments showed no gain in stored water at wheat termination compared to the amount of soil water at wheat planting. In the loamy fine sand, the wheat residue caused a reduction ($P \leq 0.05$) of 14% in soil water content. At Brownfield, the bare soil gained 71% ($P \leq 0.05$) soil water and at New Deal the gain was 37% ($P \leq 0.05$) from wheat planting to wheat termination. The statistical difference ($P \leq 0.05$) between the amount of soil water at wheat termination at Brownfield was 36 mm between the bare soil and wheat residue treatments. The statistical difference ($P \leq 0.05$) between soil water at wheat termination at New Deal was 40 mm of water in the top 0.6 m of soil. Therefore, there was no gain, 36 vs. 40 mm, in stored water by the wheat residue plots over the spring of 1995 at either location, while the bare soil was able to increase the water content in the top 0.6 m of soil by more than 30% ($P \leq 0.05$) over the spring months.

3.2. Wicking

The second objective was to determine if upward movement of water through the wheat residue, i.e., wicking, occurred after the wheat was terminated and before the cotton was planted. The amount of water lost to evaporation from both the bare soil and terminated wheat plots between wheat termination and cotton emergence for 1995 at Brownfield and New Deal are shown as a bar graph in Figure 1. These results suggested, and as expected the evaporation of water was greater in the finer textured soil when compared to the coarser textured soil. In the fine loamy sand soil, the bare soil evaporated 44% more ($P \leq 0.05$) soil water than the wheat residue. However, in the clay loam soil the wheat residue evaporated 66% more ($P \leq 0.05$) soil water than the bare soil. This result suggested that there might be a soil textural effect on the amount of water that can evaporate from a bare soil and from a terminated wheat residue [19] .

It was speculated that differences in water evaporation from the bare and from the wheat-covered soil were due to the different textures and the process of wicking. The soil at New Deal is a clay loam, i.e., more conducive for upward capillary flow of water; while Brownfield has a loamy fine sand, whose large pores would decrease capillary flow [13] . Under wicking, wheat residue could lose more water to evaporation than bare soil because many roots act as capillaries and the surface area provided by the leaves of the terminated wheat contribute to the evaporation. A possible solution to this problem could be to mechanically sever the roots of the wheat plant after termination at a 0.2 - 0.3 m depth and in this way disrupt the capillary movement of water from the soil to the atmosphere.

The cumulative evaporation of soil water from the bare soil and terminated wheat, between wheat termination to cotton emergence is shown in Figure 2. The bare soil had an 18% larger ($P \leq 0.05$) amount of water evaporation loss compared to the wheat residue. This is a result that in part may be explained by the frequent irrigation and rain events that occurred during the study. These wetting events resulted in a bare soil that was wetter than the top of the wheat residue, leading to near free-water evaporation to occur at faster rate than wicking. However, the results from this study were not conclusive to determine the effects of wicking in a clay loam soil and the measurement of E_{soil}.

Table 5. Amount of water in the top 0.6 m of the soil for wheat residue at planting and termination at two locations.

Year	Brownfield Residue Treatment [mm]				New Deal Residue Treatment [mm]			
	Bare Soil		Residue		Bare Soil		Residue	
	Planting	Termination	Planting	Termination	Planting	Termination	Planting	Termination
1995	47.0	80.5	51.5	44.5	104.0	142.5	102.0	103.0

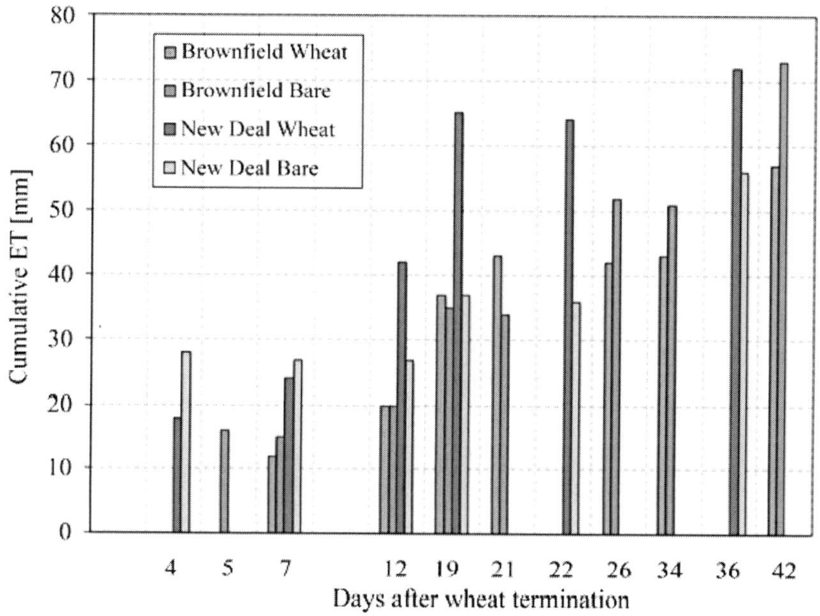

Figure 1. Cumulative measured evapotranspiration (ET, mm) as a function of days after wheat termination at Brownfield (BF) and New Deal (ND) from the wheat and bare soil residue treatment for the 1995-growing season.

Part of the difficulty in determining differences of E_{soil} under field conditions is that the amount of water that may be lost due to wicking is of the order of 1 $mm \cdot d^{-1}$ and this loss takes place from mainly the surface 0.1 m layer of the soil profile. This rate of water evaporation represents a change of 0.01-$m^3 \cdot m^{-3}$ volumetric water content on a daily basis, which is beyond the sensitivity of current instrumentation to measure volumetric soil water content [46] . Nevertheless, a wicking loss of 1 mm·d^{-1} for 20 days represents 20 mm of water, and this amount of water could determine the difference of having adequate water in the seedbed to germinate and to emerge the cotton crop planted in the residue [13] .

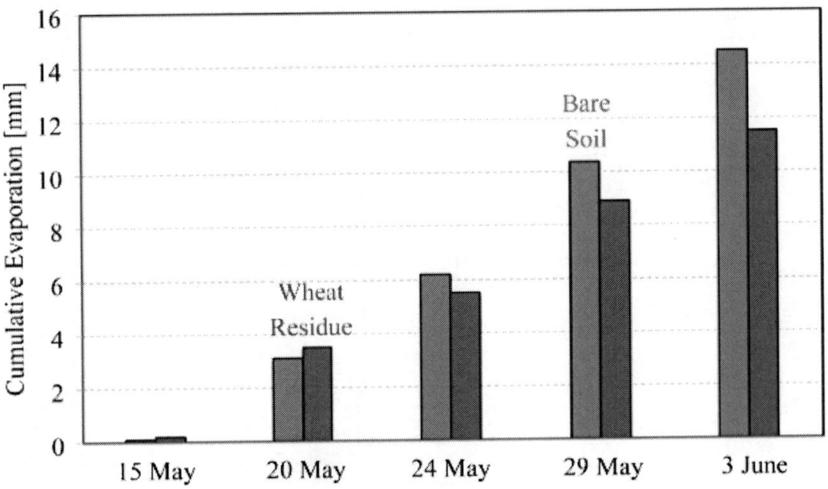

Figure 2. Cumulative measured evaporation in mm from a bare soiland wheat residue treatment for the time period between wheat termination (15 May 1996) to cotton emergence (3 June 1996) in New Deal TX.

The water used by the wheat and lost by E_{soil} affects the amount of water available for cotton growth. The amount of water content in the 1.2 m soil profile at cotton planting for both sites and growing seasons season is shown in Table 6. The bare soil treatments had more water at cotton planting than the wheat residue treatments at both locations except for the 40% ET_o plots at New Deal and the 50% ET_o plots at Brownfield for 1994. All the treatments in New Deal excluding the 40% ET_o irrigation treatment in 1994 showed a significant increase ($P \leq 0.05$) in soil water for the bare soil treatments compared to the wheat residue treatments, while Brownfield did not have a significant increase ($P \leq 0.05$) in soil water for the bare soil treatments at cotton planting. This stored water is important, particularly at the time when the cotton is planted as it necessary for seed germination and emergence impacting the vitality and vigor of the cotton seedling, which can later affect lint yields [49] [50] .

Even though the wheat used more water in Brownfield in 1995 than the bare soil (Table 5), it appears that the wheat only extracted water from the top 0.6 m of soil because the water depth in the top 1.2 m of soil at cotton planting (Table 6) showed that the soil profile for each residue treatment contained about the same amount of water in the top 1.2 m of soil. However, at New Deal, capillary movement of water by a hydraulic gradient and by wheat roots may have caused a reduction in the amount of water in the top 1.2 m of soil for the wheat residue plots, allowing the bare soil to store more water at the time that cotton was planted (Table 6).

3.3. Cotton Water Use

The last and third objective of this study was to determine the effects bare and residue-covered soils have on the water available for cotton growth and development. Water inputs and water used by the cotton in both seasons and at both sites are given in Table 7. There were no significant differences ($P \leq 0.05$) in the amount of water used by the cotton for any of the residue treatments at Brownfield (Table 7). However, the 80% ET_o bare soil treatments used at least 10% more ($P \leq 0.05$) water to grow the cotton than the residue treatments for both years at New Deal (Table 7) and as previously discussed, could be accounted for sampling error (Table 5).

Cotton lint yields were used to evaluate the effects residues have on cotton growth and development. Cotton lint yields for both locations and growing seasons are given in Table 8. These cotton lint yield values were similar to cotton yields reported for surrounding areas [51] [52] . In 1994, both locations had significantly higher ($P \leq 0.05$) lint yields (at least 8% higher) from the conventionally tilled treatments than the wheat residue plots for all irrigation treatments (Table 8). The only treatment with a loss in lint yield from the conventionally tilled treatments was in the in the low irrigation treatment at Brownfield in 1995 with a 14% decrease ($P \leq 0.05$) in lint yields. The greatest gain in lint yield from the conventionally tilled treatments over the terminated wheat treatments occurred in the 40% irrigation treatment at New Deal in 1994 with a 42% gain ($P \leq 0.05$). Therefore, our results suggested that the presence of residues had a significant ($P \leq 0.05$) impact with a loss of lint yield for both years of the study. This result confirms our hypothesis that in a semiarid region the use of a small grain cover crop under irrigated conditions uses more water than it conserves compared to conventional tilled cotton, reducing cotton lint yields.

As expected and in climates where water is a limiting factor in crop production, the irrigation treatments at each location that received the greatest amount of water resulted in higher lint yields compared to the lower irrigation amounts (Table 8). For example, Bordovsky et al. [45] showed that higher irrigation amounts also gave higher yields, but also reported that at a 50% ET_o irrigation level, conventionally tilled and wheat residue cotton systems yielded about the same ($P \leq 0.05$), while at a 100% ET_o irrigation level, the conventionally tilled cotton cropping system yielded 17% less ($P \leq 0.05$) as compared to the wheat residue treatment at Lamesa, TX, which has a sandy loam soil.

Table 6. Amount of water in the top 1.2 m of the soil when cotton was planted at two locations and two years.

Year	Brownfield Residue Treatment [mm]		New Deal Residue Treatment [mm]	
	Bare Soil	Residue	Bare Soil	Residue
1994	207.0	201.5	296.0	299.0
1995	177.5	169.5	238.0	192.0

Table 7. Water balance (inputs and outputs), for two irrigation treatments at two locations and two years.

Brownfield							
Year	Water Inputs [mm]			Water Outputs [mm]			
	Irrigation Treatment		Rain [mm]	Irrigation Treatment			
	50% ET_o	100% ET_o		50% ET_o		100% ET_o	
				Residue Treatment		Residue Treatment	
				Bare Soil	Residue	Bare Soil	Residue
1994	125	200	93	276	295	395	403
1995	41	81	375	315	314	406	403

New Deal							
Year	Water Inputs [mm]			Water Outputs [mm]			
	Irrigation Treatment		Rain [mm]	Irrigation Treatment			
	40% ET_o	80% ET_o		40% ET_o		80% ET_o	
				Residue Treatment		Residue Treatment	
				Bare Soil	Residue	Bare Soil	Residue
1994	83	140	172	283	304	442	387
1995	93	130	382	541	539	620	558

Table 8. Average cotton lint yields for two locations, and two irrigation and residue treatments, for two years.

Brownfield – Cotton Lint Yield [kg·ha^{-1}]				New Deal Cotton Lint Yield [kg·ha^{-1}]			
Irrigation Treatment				Irrigation Treatment			
50% ET_o		100% ET_o		40% ET_o		80% ET_o	
Residue Treatment		Residue Treatment		Residue Treatment		Residue Treatment	
Bare Soil	Residue	Bare Soil	Residue	Bare Soil	Residue	Bare Soil	Residue
440 (20)*	392 (16)	933 (31)	816 (11)	561 (19)	394 (23)	674 (11)	599 (21)
380 (17)	444 (21)	594 (26)	546 (24)	436 (11)	401 (31)	603 (19)	559 (11)

*The standard error of the mean is in parenthesis. Mean separation was calculated using PROC MIXED with a t-type confidence interval of 0.05.

4. CONCLUSIONS

In this study, the cotton water use and lint yield of two cotton-cropping systems in the semiarid climate of the Texas High Plains was compared. These cropping systems were conventionally tilled and cotton planted into terminated wheat, on two sites and a low and high irrigation rate over two growing seasons. Our objective was to evaluate the effectiveness of planting cotton into a crop residue obtained by terminating a winter wheat crop planted in December. The purpose of the residue is mainly to provide a mechanical barrier to wind and thus protect cotton seedlings at a time when they are most vulnerable to sandblast damage due to high wind speeds. The selected sites, allowed us to compare the impact of soil texture, i.e., coarse (Brownfield, TX) and fine (New Deal, TX). The residue treatments did not affect the total amount of water evaporated and transpired during the two seasons. However, at both locations the bare residue treatment had a higher lint yield than the cotton grown under the wheat residue. Wicking, the capillary movement of water through the wheat residue was only statistically significant at the site with the coarser texture (Brownfield, TX) and only in one (1994) of the two years.

These results suggest that the use of terminated wheat residue did not have a positive impact on increasing water storage for the subsequent cotton crop. Furthermore, terminated wheat residue adversely affected cotton lint yield on both years and sites and thus no advantages found in planting cotton into a terminated wheat residue in the semiarid Texas

High Plains. Nevertheless, over a longer period of time, the use of a wheat residue may improve soil structure, which may lead to higher infiltration rate reducing runoff and storing more rainfall.

It was postulated that in a semiarid climate growing a small grain cover crop might use more water than it conserves when combined with a cotton crop. During the 1995 wheat growth period, the conservation tillage treatment used 20% more water than the conventional tillage treatment for all treatments at both locations supporting our hypothesis. It was further hypothesized that the use of residue would not increase cotton lint yields, and all the conventionally tilled treatments for two seasons had higher lint yields except for one plot in 1994 at Brownfield over the wheat residue plots. Therefore, for these two locations for the 1994 and 1995 cotton crop seasons, the use of terminated wheat residue did not improve the growing conditions and environment for the cotton crops compared to conventional tillage cropping systems. The significance of this work is that growers using this "conservation compliance" tillage system may not be using water optimally and may even be experiencing decreased cotton lint yields.

ACKNOWLEDGEMENTS

This work is a tribute to the many research contributions of Dr. Dan Krieg, Professor emeritus, Plant and Soil Science, Texas Tech University, to crop physiology and agronomy of dryland cropping systems. On the 18 June 2015 he lost his battle with cancer.

REFERENCES

1. USDA-NASS (2014) Texas Crop Production, USDA—National Agricultural Statistics Service, Issue No. PR-163-14, 10 December 2014. http://www.nass.usda.gov/Statistics_by_State/Texas/Publications/Current_News_Release/2014_Rls/pr16314.pdf

2. Lascano, R.J. (2000) A General System to Measure and to Calculate Daily Crop Water Use. Agronomy Journal, 92, 821-832.

3. Colaizzi, P.D., Gowda, P.H., Marek, T.H. and Porter, D.O. (2009) Irrigation in the Texas High Plains: A Brief History and Potential Reductions in Demand. Irrigation & Drainage, 58, 257-274.

4. Lascano, R.J., Van Bavel, C.H.M., Hatfield, J.L. and Upchurch, D.R. (1987) Energy and Water Balance of a Sparse Crop: Simulated and Measured Soil and Crop Evaporation. Soil Science Society of America Journal, 51, 1113-1121.

5. Lascano, R.J. and Baumhardt, R.L. (1996) Effects of Crop Residue on Soil and Plant Water Evaporation in a Dryland Cotton System. Theoretical Applied Climatology, 54, 69-84.

6. Krieg, D.R. (1996) Physiological Aspects of Ultra Narrow Row Cotton Production. Proceedings of the Beltwide Cotton Conference, Nashville, TN. National Cotton Council, Memphis, 9-12 January 1996, 66.

7. Darawsheh, M.K., Khah, E.M., Aivalakis, G., Chachalis, D. and Sallaku, F. (2009) Cotton Row Spacing and Plant Density Cropping Systems. I. Effects on Accumulation and Partitioning of Dry Mass and LAI. Journal of Food, Agriculture & Environment, 7, 258-261.

8. Lascano, R.J., Baumhardt R.L., Hicks, S.K. and Heilman, J.L. (1994) Soil and Plant Water Evaporation from Strip-Tilled Cotton: Measurement and Simulation. Agronomy Journal, 86, 987-994.

9. Musick, J.T., Pringle, F.B., Harman, W.L. and Stewart B.A. (1990) Long-Term Irrigation Trends-Texas High Plains. Applied Engineering Agriculture, 6, 717-724.

10. Lascano, R.J. and Nelson, J.R. (2014) Circular Planting to Enhance Rainfall Capture in Dryland Cropping Systems at a Landscape Scale: Measurement and Simulation. In: Ahuja, L.R., Ma, L. and Lascano, R.J., Eds., Practical Applications of Agricultural System Models to Optimize the Use of Limited Water, Advances in Agricultural

Systems Modeling, ASA, CSSA, SSSA, Madison, Volume 5, 85-111.

11. Bordovsky, J.P., Lyle, W.M. and Keeling, J.W. (1994) Crop Rotation and Tillage Effects on Soil Water and Cotton Yield. Agronomy Journal, 86, 1-6.
12. Baker, J.T., McMichael, B., Burke, J.J., Gitz, D.C., Lascano, R.J. and Eprath, J.E. (2009) Sand Abrasion Injury and Biomass Partitioning in Cotton Seedlings. Agronomy Journal, 101, 1297-1303.
13. Koekkoek, E.J.W., Lascano, R.J., Hicks, S.K., Krieg, D.R. and Stroosnijder, L. (1995) Loss of Water through Terminated Wheat Plants: A Wick Effect. Proceedings of the 1995 Annual Meetings, St. Louis, 29 October-3 November 1995, 15.
14. Van Rensburg, L.D. (2010) Advances in Soil Physics: Application in Irrigation and Dryland Crop Production, South African. Journal of Plant and Soil, 27, 9-18.
15. Unger, P.W. (1994) Residue Management for Winter Wheat and Grain Sorghum Production with Limited Irrigation. Soil Science Society American Journal, 58, 537-542.
16. Nielsen, D.C., Unger, P.W. and Miller, P.R. (2005) Efficient Water Use in Dryland Cropping Systems in the Great Plains. Agronomy Journal, 97, 364-372.
17. AG-CARES (2004-2008) AG-CARES Annual Report. Texas A & M AgriLife Research & Extension Center, Lubbock. http://lubbock.tamu.edu/ag-cares/
18. Helms Farm Research Reports (2001-2005) Helms Farm Annual Reports. Texas A & M AgriLife Research & Extension Center, Lubbock. http://lubbock.tamu.edu/programs/disciplines/irrigation-water/helms-farm-research-reports/
19. Jones, O.R., Hauser, V.L. and Popham, T.W. (1994) No-Tillage Effects on Infiltration, Runoff, and Water Conservation on Dryland. Transactions of the ASAE, 37, 473-479.

20. Howell, T.A., Steiner, J.L., Schneider, A.D. and Evett, S.R. (1995) Evapotranspiration of Irrigated Winter Wheat— Southern High Plains. Transactions of the ASAE, 38, 745-759.

21. Guerif, J., Richard, G., Durr, C., Machet, J.M., Recous, S. and Roger-Estrade, J. (2001) A Review of Tillage Effects on Crop Residue Management, Seedbed Conditions and Seedling Establishment. Soil & Tillage Research, 61, 13-22.

22. Hobbs, P.R., Sayre, K. and Gupta, R. (2008) The Role of Conservation Agriculture in Sustainable Agriculture. Philosophical Transactions of the Royal Society B: Biological Sciences, 363, 543-555.

23. Alvarez, R. and Steinbach, H.S. (2009) A Review of the Effects of Tillage Systems on Some Soil Physical Properties, Water Content, Nitrate Availability and Crops Yield in the Argentine Pampas. Soil & Tillage Research, 104, 1-15.

24. Baumhardt, R.L., Keeling, J.W. and Wendt, C.W. (1993) Tillage and Residue Effects on Infiltration into Soils Cropped to Cotton. Agronomy Journal, 85, 379-383.

25. Dao, T.H. (1993) Tillage and Winter Wheat Residue Management Effects on Water Infiltration and Storage. Soil Science Society of America Journal, 57, 1586-1595.

26. Hoogmoed, W.B. and Stroosnijder, L. (1984) Crust Formation on Sandy Soils in the Sahel. I. Rainfall and Infiltration. Soil & Tillage Research, 4, 5-23.

27. Lascano, R.J., Vorheis, J.T., Baumhardt, R.L. and Salisbury, D.R. (1997) Computer-Controlled Variable Intensity Rain Simulator. Soil Science Society of America Journal, 61, 1182-1189.

28. Alberts, E.E. and Neibling, W.H. (1994) Influence of Crop Residues on Water Erosion. In: Unger, P.W., Ed., Managing Agricultural Residues, Lewis Publ., Chelsea, MI, 19-39.

29. Savabi, M.R. and Stott, D.E. (1994) Plant Residue Impact on Rain Interception. Transactions of the ASAE, 37, 1093-1098.

30. Steiner, J.L. (1994) Crop Residue Effects on Water Conservation. In: Unger, P.W., Ed., Managing Agricultural Residues, Lewis Publ., Chelsea, MI, 41-76.

31. Baumhardt, R.L., Schwartz, R., Howell, T.A., Evett, S.R. and Colaizzi, P. (2013) Residue Management Effects on Water Use and Yield of Deficit Irrigated Cotton. Agronomy Journal, 105, 1026-1034.

32. Lascano, R.J. and Van Bavel, C.H.M. (1986) Simulation and Measurement of Evaporation from a Bare Soil. Soil Science Society of America Journal, 50, 1127-1132.

33. Farahani, H.J. and Ahuja, L.R. (1996) Evapotranspiration Modeling of Partial Canopy/Residue-Covered Fields. Transactions of the ASAE, 39, 2051-2064.

34. Klocke, N.L., Currie, R.S. and Aiken, R.M. (2009) Soil Water Evaporation and Crop Residues. Transactions of the ASABE, 52, 103-110.

35. Steiner, J.L. (1989) Tillage and Surface Residue Effects on Evaporation from Soils. Soil Science Society American Journal, 53, 911-916.

36. Enz, J.W., Brun, L.J. and Larsen, J.K. (1988) Evaporation and Energy Balance for Bare and Stubble Covered Soil. Agricultural and Forest Meteorology, 43, 59-70.

37. Horton, R., Kluitenberg, G.J. and Bristow, K.L. (1994) Surface Crop Residue Effects on the Soil Surface Energy Balance. In: Unger, P.W., Ed., Managing Agricultural Residues, Lewis Publ., Chelsea, MI, 143-162.

38. Heilman, J.L., McInnes, K.J., Gesch, R.W. and Lascano, R.J. (1992) Evaporation from Ridge-Tilled Soil Covered with Herbicide-Killed Winter Wheat. Soil Science Society American Journal, 56, 1278-1286.

39. Fryrear, D.W. and Bilbro, J.D. (1994) Wind Erosion Control with Residues and Related Practices. In: Unger, P.W., Ed., Managing Agricultural Residues, Lewis Publ., Chelsea, MI, 7-17.

40. Evett, S.R. and Lascano, R.J. (1993) ENWATBAL.BAS: A Mechanistic Evapotranspiration Model Written in Compiled Basic. Agronomy Journal, 85, 763-772.

41. Vorheis, J.T. (1997) Water Balance of Cotton Cropping Systems. Master's Thesis, Plant and Soil Science Department, Texas Tech University, Lubbock, 38 p.

42. Ralston, J.T. (1997) Management Strategies for Dryland Cotton Production in West Texas. Master's Thesis, Plant and Soil Science Department, Texas Tech University, Lubbock, 47 p.

43. Lyle, W.M. and Bordovsky, J.P. (1981) Low Energy Precision Application (LEPA) Irrigation System. Transactions of the ASAE, 24, 1241-1245.

44. Allen, R.G., Walter, I.A., Elliot, R., Howell, T.A., Itenfisu, D. and Jensen, M.E. (2005) The ASCE Standardized Reference Evapotranspiration Equation. ASCE-EWRI Task Committee Report, January 2005, 70 p. http://www.kimberly.uidaho.edu/water/asceewri/ascestzdetmain2005.pdf

45. Bordovsky, J.P., Lyle, W.M., Lascano, R.J. and Upchurch, D.R. (1992) Cotton Irrigation Management with LEPA Systems. Transactions of the ASAE, 35, 879-884.

46. Evett, S.R. (2007) Soil Water and Monitoring Technology. In: Lascano, R.J. and Sojka, R.E., Eds., Irrigation of Agricultural Crops, 2nd Edition, ASA, CSSA, SSSA, Madison, 25-84.

47. Baumhardt, R.L., Lascano, R.J. and Krieg, D.R. (1995) Physical and Hydraulic Properties of a Pullman and Amarillo Soil on the Texas South Plains. Technical Report D.R. No. 95-1, Texas A & M University Agricultural Research and Extension Center, Lubbock/Halfway.

48. Bertuzzi, P., Bruckler, L., Bay, D. and Chanzy, A. (1994) Sampling Strategies for Soil Water Content to Estimate evapotranspiration. Irrigation Science, 14, 105-115.

49. Peng, S., Krieg, D.R. and Hicks, S.K. (1989) Cotton Lint Yield Response to Accumulated Heat Units and Soil Water Supply. Field Crop Research, 19, 253-262.

50. Morrow, M.R. and Krieg, D.R. (1990) Cotton Management Strategies for a Short-Growing Season Environment: Water-Nitrogen Considerations. Agronomy Journal, 92, 52-56.

51. USDA-NASS (1994) Agricultural Statistics-1994. http://www.nass.usda.gov/Publications/Ag_Statistics/agr4all.pdf

52. USDA-NASS (1995) Agricultural Statistics-1995. Chapter II—Statistics of Cotton, Tobacco, Sugar Crops, and Honey. http://www.nass.usda.gov/Publications/Ag_Statistics/1995-1996/agr95_2.pdf

NOTES

*The US Department of Agriculture (USDA) prohibits discrimination in all its programs and activities on the basis of race, color, national origin, age, disability, and where applicable, sex, marital status, familial status, parental status, religion, sexual orientation, genetic information, political beliefs, reprisal, or because all or part of an individual's income is derived from any public assistance program.

#Mention of this or other proprietary products is for the convenience of the readers only and does not constitute endorsement or preferential treatment of these products by the USDA-ARS.

CHAPTER 3

An Integrated Geochemical and Mineralogical Approach for the Evaluation of Zn Distribution in Long-Term Sludge-Amended Soil

Dominique Proust

LIENSs UMR 7266 CNRS, La Rochelle, France

ABSTRACT

This research work was designed to compare the Zn distribution in a long-term sludge-amended soil with that in a control soil. Two complementary approaches were performed: 1) a geochemical approach at the metric scale of the bulk soil horizons and 2) a mineralogical approach at the micrometric scale of the primary minerals weathering microsites. The geochemical approach revealed that Zn in the control soil was inherited from the weathering parent-rock. Its concentration was always lower than in the amended soil where Zn was supplied at the surface by the spread sludges and moves downwards. The mineralogical approach showed that the clay minerals, produced by the weathering of the primary minerals (amphiboles and plagioclases), or filling the fissure network were made up of smectites (saponite and montmorillonite) at the bottom and kaolinite at the top of the two soil profiles. Each clay mineral, with its specific sorption capacity, controlled the Zn distribution within the soil: the smectites produced by the amphiboles had high sorption capacity and favored Zn retention in the upper horizons of the soil. Conversely, the kaolinites produced by the plagioclases had lower sorption capacity, did not retain Zn in the surface horizons, and allowed it to migrate to deeper horizons where it was sorbed onto the montmorillonites.

KEYWORDS

Weathering, Soil, Clay Minerals, Heavy Metal, Zinc

1. INTRODUCTION

The spreading of sewage sludges onto farmlands can be a beneficial method for soil amendment because the sludges can be a source of plant nutrients (especially N, P) and organic matter. However, much of these sludges results from the treatment of industrial and urban by-products and may contain toxic organic and inorganic compounds. Among the inorganic components, the heavy metals have been critically examined since some of them can be toxic to the biosphere at very low concentrations: for instance, Cd concentration cannot exceed 2 $\mu g \cdot g^{-1}$ for soils to be selected for sludge spreading [1] . The origin of the heavy metals in a soil is either geogenic if they are inherited from the weathering of the parent-rock [2] -[4] , or anthropogenic if they come, among other sources, from contamination by repeated application of inorganic fertilizers [5] , sludge spreading [6] , and/or diffuse atmospheric contamination [7] . One of the most valuable properties of soil as regards sludge-spreading is its ability to absorb and retain the heavy metals ions. The heavy metals adsorption/desorption reactions occur at the soil solid/solution interface and involve especially the clay minerals which are the essential reactive solid components in soils. These clay minerals, with their natural sorptive capacities for heavy metals, are of particular relevance for sludge spreading and have received considerable attention in the last decades [8] -[10] .

The adsorption of heavy metals onto clay minerals operates in particular sites with permanent or variable negative charges [11] [12] . The sites with permanent charges are located at the interlayer basal surfaces of the clay minerals and result from the isomorphic substitutions of Al^{3+} for Si^{4+} in the silica tetrahedral sheet and/or Fe^{2+}/Mg^{2+} for Al^{3+} in the alumina octahedral sheet. These interlayer negative charges are neutralized by pH-in- dependent adsorption of cations as outer-sphere complexes (cation exchange reactions). The sites with variable charges are located at the crystal edge surfaces and originate in the protonation or deprotonation of surface hydroxyl (SOH) groups. These sites promote pH-dependent adsorption attributed to surface complexation reactions with silanol and aluminol groups to form inner-sphere complexes.

The major part of the clay minerals found in the soils results from the weathering of the parent-rock in two simultaneous processes: i) the dissolution of the primary rock-forming minerals with the release of the cations and the heavy metals; and ii) the crystallization of clay minerals from the soil solutions with the sorption of the previously released heavy metals. Mineralogical studies have shown that the chemical weathering reactions and their associated metals release are active in specific soil microsystems with their own solid and solution chemical properties [13] - [15] . They are typically micrometric composite aggregates of clay minerals (specific to the primary mineral being weathered), oxihydroxides, and organic matter moistened with soil solutions.

Up to now, studies dealing with polluted soils described heavy metals contaminations at the soil scale, i.e. in the fine earth (<2 mm) of each soil horizon [16] [17] and/or in separated soil fractions, e.g. <2 μm clay fraction or 2 - 50 μm silt fraction [18] [19] . These conventional pedological approaches, however, underestimate the influence of the individual soil components, especially the weathering microsystems, upon the adsorption and retention of heavy metals in soils.

Even though the bulk concentrations of heavy metals can be decisive in assessing global soil pollution, they result from the addition of the partial heavy metals concentrations in each of the weathering microsystems. As a consequence, the detailed analysis of a soil contaminated with heavy metals requires both the knowledge of its bulk metals concentration but also the partial metals contents in the microsystems in which they are retained. By taking advantage of a long-term spreading on soil of highly Zn concentrated sewage sludges (498 mg · kg^{-1}), this study is undertaken to estimate i) the impact of the sludge application upon Zn distribution and migration in the soil and ii) the influence of the clay mineral species upon Zn retention in the weathering microsystems.

2. MATERIALS AND METHODS

2.1. Soil Characteristics

The soils investigated in this study were selected and sampled during the geological and pedological field survey of the "Saint-jean-Ligoure" diorite massif in Haute-Vienne (Limousin, France), 25 km in the South of Limoges (Figure 1). They are inceptisols where wet sewage sludges were deposited during the 90's, each year for ten years, at an application rate of 2500 kg · ha^{-1} · $year^{-1}$ with mean Zn concentration of 498 mg · kg^{-1}. Two soils were sampled on the same dioritic parent-rock: the Amended Soil (AS) that received the sewage sludges for ten years and the Control Soil (CS) that is free of sludge spreading.

The two soils are moderately drained, deep with the weathered rock found at 100 - 150 cm depth and show the typical A, Bw and C horizons sequence of inceptisols (Figure 2(a), Figure 2(b)). The R horizon, i.e. the coherent unweathered parent rock (Figure 2(c)), is a diorite which appears to the naked eye as a "salt and pepper" coarse-grained rock made up of major amphiboles, plagioclases, and minor quartz. The two soils have very similar physicochemical properties (Figure 2(d)) with slightly acidic to almost neutral pH in all horizons (6.1 - 6.8). The pH values increase with depth, suggesting some leaching of base cations or probably the effect of organic matter with 1.82% to 1.95 % organic carbon content in the A horizons, decreasing to 0.85% - 0.63 % in the Bw horizon and disappearing in the C horizon. The cation exchange capacities of the two soils are similar with medium values ranging between 26.8 and 31.9 cmol · kg^{-1}, whereas the clay contents, typical of the clay loam texture observed in the inceptisols, increase slightly in the Bw horizon (cambic horizon).

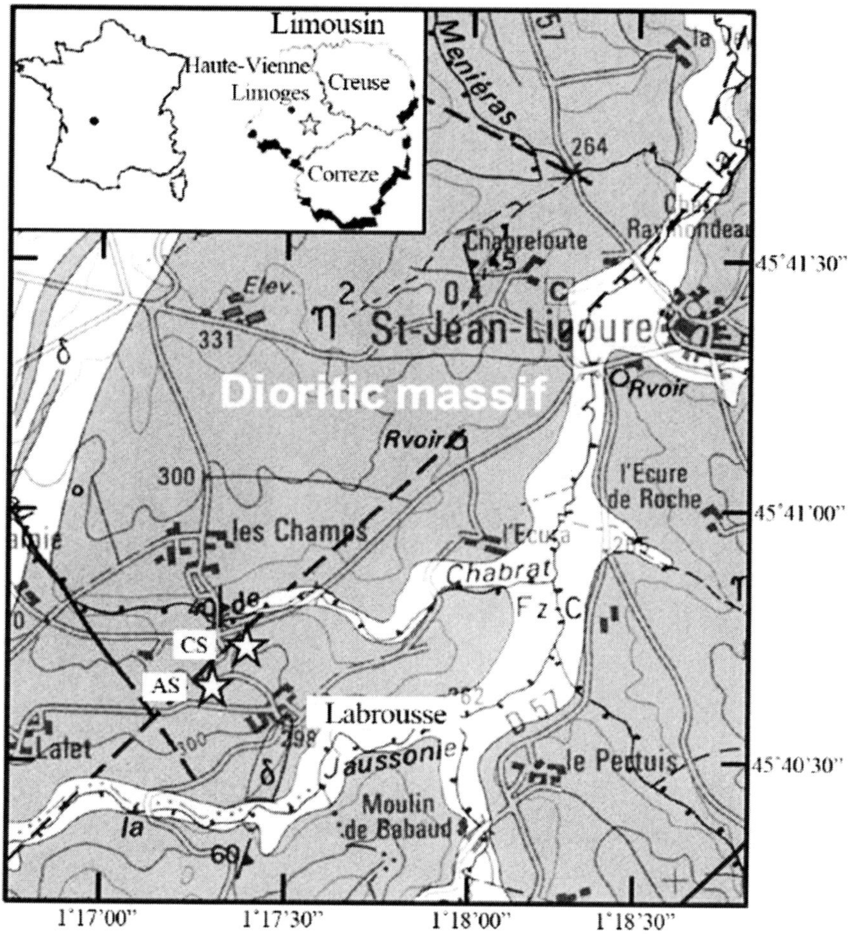

Figure 1. Geological map of the sudied area in the Limousin region, France, and location of the sampled soil profiles; CS:control soil, AS: Amended Soil.

2.2. Sampling Methods

Each soil was carefully sampled using plastic cylindrical corers to prevent the soil metal contamination. The cylinder, 10 cm long with a diameter of 10 cm, had cutting edges. It was smoothly and continuously pushed into the soil to collect undisturbed blocks which retained the original soil structure. Each core was divided into two subsamples. The first was disaggregated by gentle shaking in water, sieved to 2 mm, air dried and reserved for the mineral grains separation, the bulk chemical and the X-ray diffraction (XRD) analyses.

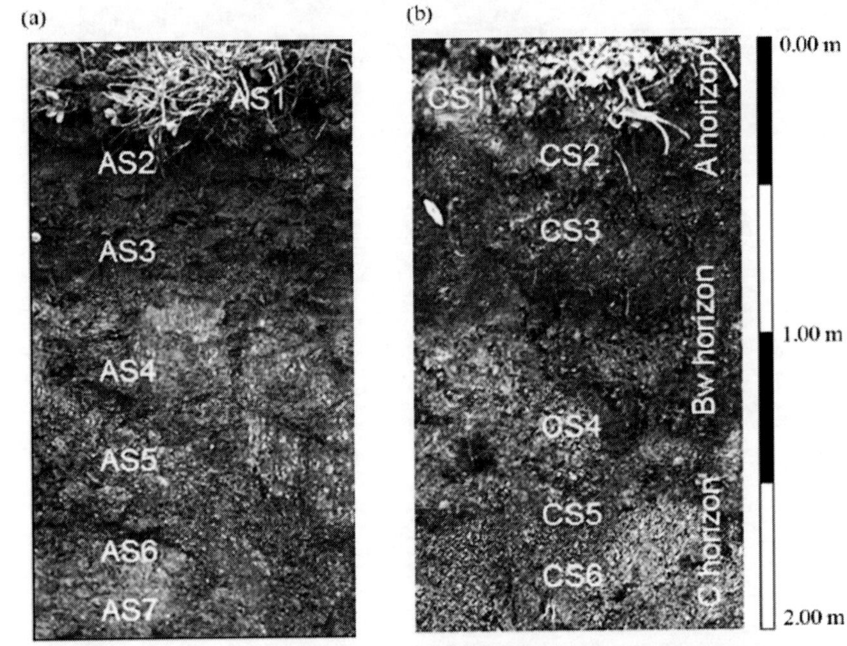

(d)

Soils	CS	AS	CS	AS	CS	AS
Horizons	A	A	Bw	Bw	C	C
pH water	6.10	6.50	6.40	6.80	6.50	6.80
Org. C%	1.92	1.82	0.63	0.85		
C.E.C.*	31.90	26.90	30.80	27.90	27.50	26.80
Clay %	27.90	29.70	28.70	31.80	25.20	24.20

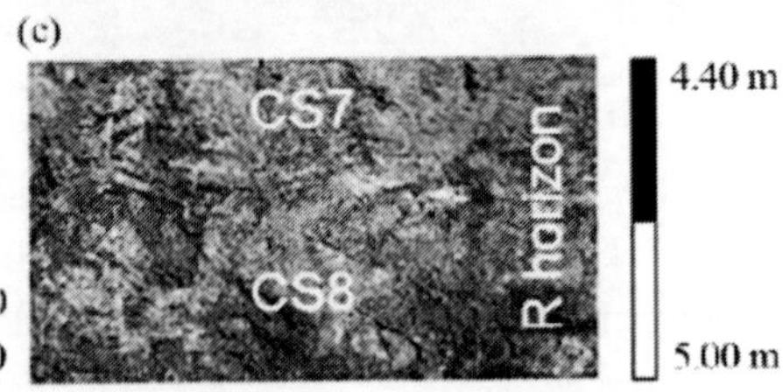

Figure 2. Sketch of the soil profiles; (a) Amended Soil (AS); (b) Control Soil (CS); (c) Parent-rock (R horizon); (d) Bulk physico-chemical properties of soils.

The second subsample was preserved undisturbed for thin sections preparation, electron probe microanalyses (EPMA) and scanning electron microscope (SEM) study.

2.3. Mineralogical Separation

The organic matter in the first <2 mm subsample was removed using 33% hydrogen peroxide. 200 grams of this sample were shaken in water to destroy the aggregates, sieved into the sandy fractions 50 - 100 μm, 100 - 150 μm, 150 - 250 μm, 250 - 500 μm, and dried at 50°C for 24 h. The observation of each sieved fraction under optical microscopy (OM) revealed that most euhedral monomineralic grains were found in the 250 - 500 μm fraction. The smaller sized fractions were made up of broken primary minerals. About 500 monomineralic grains, 250 to 500 μm in size, of amphiboles and plagioclases (the two most important minerals in the parent-rock and soils) were then needle-sorted under the stereomicroscope for the study of their specific weathering products. Separation efficiency was checked by XRD.

2.4. Bulk Chemical Analyses

The bulk chemical analyses were obtained from the first subsample, freed of its organic matter. An aliquot quantity of matter (5 g) was collected by quartering with a riffle splitter and crushed down to less than 50 μm size into agate mortar. 300 mg were taken from the total powder, fused with $LiBO_2$ at 1050°C and dissolved in 1N HNO_3 for the bulk chemical analyses. The major and trace elements were analysed using ICP-AES and ICP-MS at the Service d'Analyse des Roches et des Minéraux, (SARM), CRPG-CNRS (Vendoeuvre-lès-Nancy, France). The major and trace elements concentrations were expressed in wt.% and $mg \cdot kg^{-1}$, respectively.

2.5. X-Ray Diffraction Analyses

The bulk clay fractions, together with the clay minerals extracted from the weathered monomineralic grains by sonification, were analysed by XRD on Ca-saturated oriented preparations, air dried and glycolated. The diffractometer was a PHILIPS PW 1730 (40 kV, 40 mA) with a Fe-filtered CoKa radiation and a stepping motor driven with a DACO-MP recorder and the Diffrac-AT software (Socabim, Munich, Germany). The bulk clay fraction <2 μm was extracted from 10 g of the <2 mm first subsample. The 10 grams sampled were first dispersed in 350 ml deionized water, sonicated for 2 min at 300 W/20 kHz, then mechanically shaken for 3 h. The stable suspension was then centrifugated at 20°C and 1000 rpm for 2 mn 30 s with a Jouan JR4.22 centrifuge in order to separate the <2 μm fraction. The effectiveness of this clay separation was then checked using a laser diffraction granulometer IP Malvern Mastersizer.

2.6. Microanalyses of the Soil Samples

The second subsample was devoted to the EPMA and the SEM studies. Thin sections were first prepared according to the procedure of Camuti and McGuire [20] . Samples were hardened under vacuum (100 mbars) with an epoxy resin (ARALDITE 2020) mixed with 20% acetone thinner. After polymerization, samples were cut and thin sections obtained through polishing with silicon carbide (17 μm and 9 μm particle sizes) and diamond calibrated powders (6 μm, 3 μm, 1 μm, and 0.25 μm particle sizes).

Representative pedofeatures (clay coatings, clay cutans, etc.) and weathering microsites were first located on the thin sections, marked with black circle under an optical microscope and then "in situ" analysed for major and trace elements. EPMA were obtained using a CAMECA SX 50 electron microprobe (Service CAMPARIS, Université Paris VI) equipped with wavelength-dispersive spectrometers (WDS). The microprobe was calibrated using synthetic and natural oxides. Corrections were made with a ZAF program. A specific trace program with the electron microprobe was developed [21] to analyse heavy metals in small volumes with detection limits closed to 6 - 8 mg · kg^{-1}. Major elements analyses were performed at 15 kV and 4 nA prior to trace element analysis performed at higher voltage (30 kV) and higher beam current (100 nA and 500 nA for small and large minerals, respectively) in order to improve detection limits. The spot sizes varied from 1 to 5 - 10 μm to limit beam damage. The counting time was 10 s per major element and 1000s per trace element. Major and trace elements are expressed respectively in wt% and mg · kg^{-1}.

For each dominant primary rock forming mineral, i.e. amphibole and plagioclase, ten grains were selected from the 250 - 500 μm needle-sorted fraction and affixed to glass plates with double-sided adhesive tape. The two glass plates were then gold-coated for the micromorphological SEM study of the weathering products.

2.7. Geochemical Mass Balances

The geochemical mass balances were obtained from the calculation of enrichment factors (EF) using the method of Hernandez et al. [7] and Sterckeman et al. [4] with the equation:

$$EF = (X/R)_{altered} / (X/R)_{reference} \quad (1)$$

where the content of element X is normalized to the content of the reference element R. A value of 1 for EF indicates no element enrichment or depletion. EF values >1 indicate element enrichment and when <1 indicate element depletion. The aluminium, with its narrow range of variation in the two soils (17.53% to 20.08% in Table 1) was chosen as the reference element. The CS8 sample with the lowest loss on ignition suffered very low weathering and was selected as the reference rock for major elements and Zn bulk analyses.

Table 1. Chemical analyses of major elements (wt%) and Zn ($mg \cdot kg^{-1}$) in control soil (CS) and amended soil (AS) samples.

Sample	CS8	CS7	CS6	CS5	CS4	CS3	CS2	CS1	AS7	AS6	AS5	AS4	AS3	AS2	AS1
Depth (cm)	485	450	180	160	130	65	40	20	190	170	140	110	70	40	20
SiO_2	48.79	49.00	47.99	47.53	47.15	49.12	47.48	46.07	51.2	50.21	51.27	52.34	50.83	51.23	50.95
TiO_2	1.08	1.32	1.14	1.32	1.33	1.30	1.22	1.37	1.28	1.24	1.27	1.31	1.42	1.48	1.51
Al_2O_3	18.78	18.25	19.10	19.37	19.22	19.84	20.08	18.98	18.76	18.47	18.41	17.93	18.67	17.53	18.74
Fe_2O_3[a]	9.58	10.04	10.03	10.08	10.29	9.41	10.13	9.75	8.87	9.53	8.99	9.47	9.26	9.65	9.13
MnO	0.17	0.21	0.18	0.19	0.21	0.20	0.17	0.16	0.14	0.15	0.15	0.12	0.11	0.12	0.11
MgO	4.47	4.55	4.20	4.06	3.89	3.88	4.09	3.52	3.25	3.44	3.19	3.29	3.36	3.36	3.08
CaO	7.52	6.97	6.97	7.05	6.69	6.19	4.15	6.26	5.7	5.83	5.24	2.11	3.55	2.41	4.26
Na_2O	3.54	3.29	3.36	3.13	2.94	3.25	3.88	2.82	3.19	3.08	3.19	3.5	2.96	3.18	3.25
K_2O	1.44	1.40	1.50	1.19	1.37	1.27	1.51	1.47	1.54	1.35	1.69	1.68	1.66	1.64	1.63
P_2O_5	0.59	0.70	0.60	0.66	0.70	0.67	0.47	0.57	0.00	0.00	0.00	0.00	0.00	0.00	0.00
LOI[b]	3.77	3.94	4.68	5.21	5.04	4.62	6.50	8.53	5.91	6.48	6.19	7.77	7.74	8.49	6.72
Total	99.73	99.67	99.75	99.80	98.85	99.73	99.62	99.50	99.84	99.78	99.59	99.52	99.56	99.09	99.38
Zn	111.18	124.46	113.06	119.32	126.08	120.42	108.24	115.56	115.36	103.04	80.64	72.15	79.64	68.27	69.29

[a]Total iron expressed as Fe_2O_3. [b]LOI: Loss on ignition.

3. RESULTS

3.1. Parent-Rock Mineralogy

The two soils are derived from the same dioritic parent rock which outcrops at the bottom of the quarry exposing the control soil profile (R horizon in Figure 2(c)). The diorite appears to the naked eye as a "salt and pepper" coarse-grained rock made up of major dark-green amphiboles, white plagioclases and minor vitreous quartz. The modal rock composition was obtained from the sample CS8, i.e. the unweathered coherent rock with high apparent density ($2.92 g \cdot cm^{-3}$) and the lowest loss on ignition (3.77% in Table 1). The counting of 3000 points on thin section gives the mineralogical composition of 38% amphibole, 32% plagioclase, 12% orthoclase and minor quartz (8%), albite (7%), and titanomagnétite (3%). The EPMA of the major elements in the rock forming minerals show that the amphiboles are calcic hornblendes and plagioclases are andesine with 48% anorthite content. The EPMA of Zn (Table 2) reveal that the amphiboles are the major Zn-bearing minerals with high Zn concentrations of 180 mg · kg^{-1} versus lower Zn contents in the plagioclases (17 mg · kg^{-1}), orthoclase (20 mg · kg^{-1}) and albite (14 mg · kg^{-1}).

3.2. Geochemical Mass Balances

The enrichment factors calculated using the Equation (1) and Table 1 are plotted versus depth inFigure 3. The chemical evolution trends from rock to saprolite and soil appear similar for the major elements in the control and amended soils. The first chemical change observed during weathering is the general increase in EF calculated for the loss on ignition (EF-L.O.I.) from bottom to top of the two soil profiles (Figure 3(a)). EF-L.O.I. increases more than twice from the sample reference CS8 (EF-L.O.I. = 1) to the top of the soils (EF-L.O.I. reaches 2.41 in the sample AS2 at the top of the amended soil). This trend reflects the general increase in the bulk water content (loss on ignition at 1050°C, i.e. the hydrate content in the absence of the CO_2 found in carbonates and after organic matter destruction), and is related to the increase in the hydrated clay minerals crystallizations as weathering proceeds. In the same way as aluminium that was chosen for reference element, the measured SiO_2 and Fe_2O_3 concentrations (Table 1) remain stable throughout the two soil profiles and give EF values close to 1 (0.91 - 1.12 for EF-SiO_2 and 0.93 - 1.08 for EF-Fe_2O_3). Conversely, the alkaline earth elements Mg and Ca decrease regularly from the bottom to the top of the two soils: The EF-MgO and EF-CaO reach respectively 0.69 and 0.34 in the sample AS2 at the top of the amended soil (Figure 3(b)). This regular and general decrease in the two soils is indicative of the weathering processes which affect the amphiboles (EF-MgO and EF-CaO) and the plagioclases (EF-CaO) in the saprolite and soil.

The chemical analyses of Zn in the bulk soil horizons (Table 1) reveal that, although the sewage sludges spread on the amended soil were heavily loaded with Zn (498 mg · kg^{-1} as mean concentration), its concentration in the two soils never exceeds the threshold value for anthropogenic contamination (300 mg · kg^{-1}, [22]). The Zn enrichment factors, when calculated from Table 1 and plotted versus depth in Figure

3(c), reveals two different trends depending on whether the control soil or the amended soil is considered: The distribution of EF-Zn in the control soil does not show any significant variation throughout the soil profile, whereas EFs in the amended soil indicate pronounced depletion of Zn in the surface horizons that may suggest migration of this metal to depth (down to 190 cm, sample AS7). This Zn migration to depth has been already observed by Scokart et al. [23] .

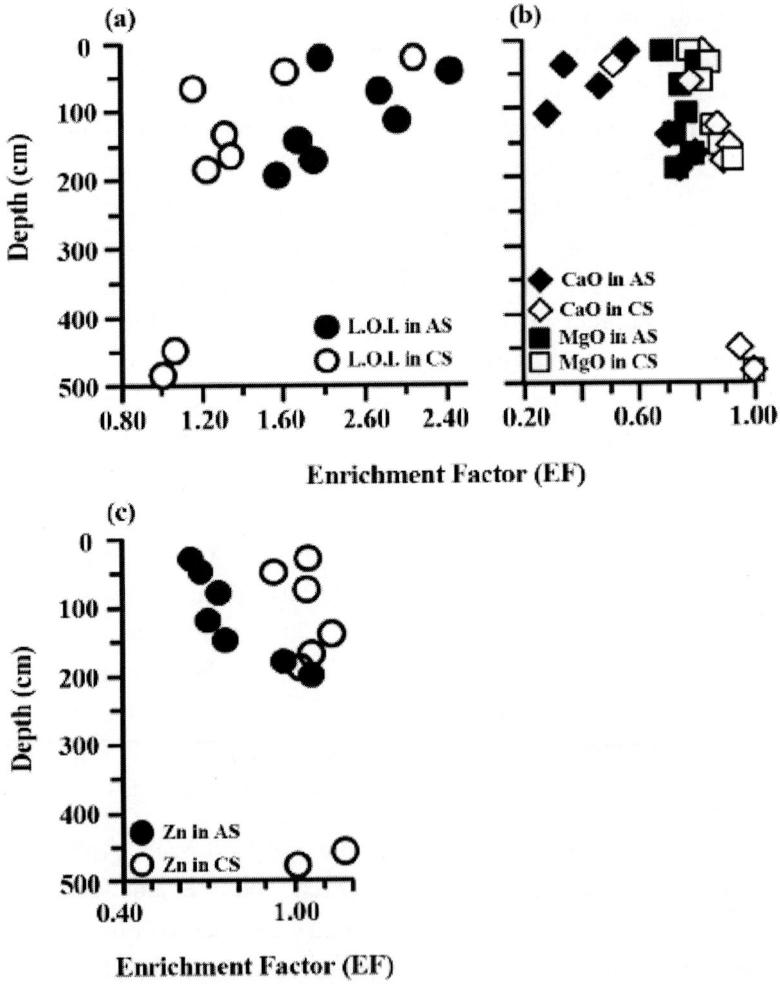

Figure 3. Variations in the enrichment factor (EF) calculated for (a) Loss on Ignition (L.O.I.); (b) CaO and MgO; (c) Zn in the amended (AS) and control (CS) soils.

Table 2. Mean EPMA of Zn (nanalyses, mg · kg^{-1}) in the rock-forming minerals.

	Amphibole		Plagioclase		Orthoclase		Albite	
	$n = 14^{a}$	SE^{b}	$n = 10$	SE	$n = 10$	SE	$n = 10$	SE
Zn	180	4	17	2	20	2	14	6

[a]Each analysis is already the mean of ten measurements. [b]Standard error.

3.3. Behaviour of Zn in the Soil Clay Fractions

The clay minerals identified in the <2 μm granulometric fractions are similar in the control and amended soils. The XRD on Ca-oriented preparations reveals the typical 001 reflections of smectite, at 15.10 Å in air-dried state and shifting to 17.05 Å after ethylene-glycol solvation. The smectite is associated with a kaolinite which is typified, in air-dried and ethylene-glycol solvated states, by 001 and 002 reflections at 7.15 Å and 3.55 Å respectively. These reflections disappear after 450°C heating. No reflection of iron oxide or hydroxide was detected on the XRD patterns of the random powders.

The Zn contents in the soil clay fractions were measured using the electron microprobe and are given with their mean standard errors in Table 3. The EPMA are plotted versus depth in Figure 4(a) to follow the Zn distribution in the two soil profiles. It appears in Table 3 that the Zn concentrations in the bulk clay fractions of the amended soil range from 144 to 203 mg · kg^{-1} and are always higher than in the control soil where they range from 10 to 24 mg · kg^{-1}. The shape of the Zn distribution profile in the amended soil (Figure 4(a)) is typically that of a downward migrating element, with Zn concentrations about ten times that measured in the control soil and a regular decrease from the top to the bottom of the amended soil. The highest Zn concentrations are measured in the surface horizons, from 195 to 203 mg · kg^{-1} at 20 cm (sample AS1) and 40 cm (sample AS2) depths, decreasing to 150 mg · kg^{-1} at the bottom of the soil (190 cm depth, sample (AS7)). For comparison, the Zn distribution profile of the control soil does not show significant variations when standard errors are considered (Figure 4(a)). The Zn concentrations are much lower than those measured in the amended soil, and range from 10 to 24 mg · kg^{-1} at 20 cm (sample CS1) and 40 cm (sample CS2) depths, and from 14 to 15 mg · kg^{-1} at 180 cm (sample CS6) and 485 cm (sample CS8) depths.

Table 3. Mean EPMA of Zn (n analyses, mg · kg^{-1}) in the clay fractions of amended (AS) and control (CS) soils.

	AS7		AS6		AS5		AS4		AS3		AS2		AS1			
Depth (cm)	190		170		140		110		70		40		20			
	$n = 4$[a]	SE[b]	$n = 4$	SE	$n = 4$	SE	$n = 4$	SE	$n = 4$	SE	$n = 4$	SE	$n = 4$	SE		
Zn	150	12	158	12	144	11	153	10	177	9	203	13	195	12		
Sample	CS8		CS7		CS6		CS5		CS4		CS3		CS2		CS1	
Depth (cm)	485		450		180		160		130		65		40		20	
	$n = 4$	SE	$n = 4$	SE	$n = 4$	SE	$n = 4$	SE	$n = 4$	SE	$n = 4$	SE	$n = 4$	SE	$n = 4$	SE
Zn	15	2	14	1	14	2	17	2	18	3	19	3	24	5	10	3

[a]Each analysis is already the mean of ten measurements. [b]Standard error.

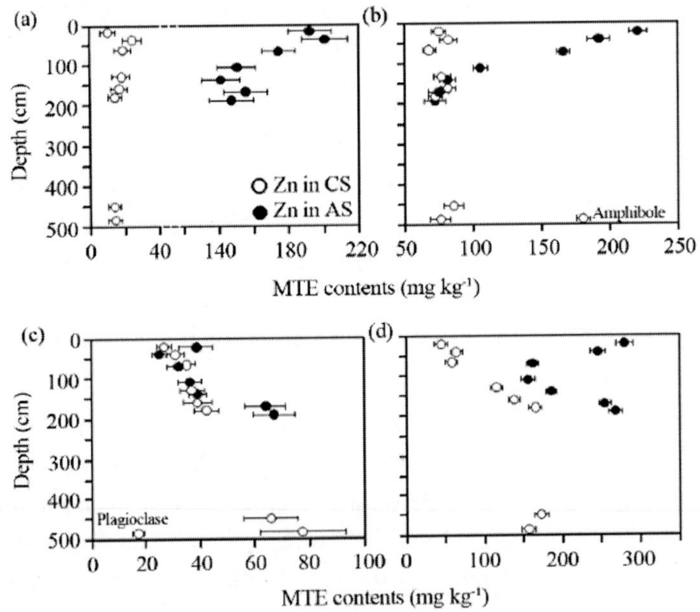

Figure 4. Vertical distribution profile of Zn in the amended (AS) and control (CS) soils; (a) Zn concentrations in the <2 μm clay fraction; (b) Zn concentrations in the amphiboles and their clay minerals weathering products; (c) Zn concentrations in the plagioclases and their clay minerals weathering products; (d) Zn concentrations in the fissural clay minerals.

3.4. Behaviour of Zn in the Amphibole Weathering Microsystems

The first signs of amphiboles weathering are found in the C horizons of the two soils. The OM observation of the thin sections reveals that euhedral amphibole prismatic crystals begin to weather along the enlarged cleavage planes (Figure 5(a)) where SEM shows the crystallization of clay minerals with the typical honeycomb texture of the smectite (Figure 5(b)). These characteristic features are commonly observed in the weathered rocks and soils [24] [25] . The weathering intensity increases in the Bw horizons where amphibole crystals appears highly fragmented into 20 - 100 μm organized residues by the opening of the intramineral microcracks, up to 50 μm wide (Figure 5(c)). The SEM reveals that the clay minerals filling these microcracks exhibit the same honeycomb morphology of smectite clay minerals, already observed in the C horizons; they are covering the saw-tooth terminations of the dissolving amphibole crystals (Figure 5(d)). The most advanced weathering stages are found in the surface A horizons where the amphiboles crystals are almost entirely replaced by a clayey weathering plasma (Figure 5(e)) made up of smectites with the characteristic honeycomb texture (Figure 5(f)).

The clay minerals, extracted from the amphiboles grains by sonication, were identified by XRD. The XRD patterns are similar in the C, Bw, and A horizons with the same typical reflections of smectite. The Ca-oriented preparations reveals strong 001 reflections ranging from 14.50 Å to 15.49 Å in air-dried state, shifting to 16.87 - 17.14 Å with higher orders at 8.46 Å and 3.40 Å after ethylene-glycol solvation. This expansion to 17 Å with ethylene-glycol is typical of smectite layers.

The EPMA of the amphibole weathering products characterize two chemically distinct smectite populations. The first is detected in the C horizons of soils and is made up of magnesian smectites belonging to the trioctahedral group of saponites. The second population is found in more advanced weathering stages (Bw and A horizons) as aluminous smectites belonging to the dioctahedral smectite group of montmorillonites.

The Zn contents in the clay minerals produced by the weathering of amphiboles were measured using the electron microprobe and are given in Table 4 with their mean standard errors. The EPMA are plotted versus depth in Figure 4(b) to follow the behaviour of Zn in the two soil profiles. The early weathering of amphiboles into saponites (R horizon, samples CS8 and CS7) results into the partial release of Zn, from 180 mg · kg^{-1} in the fresh amphibole (Table 2) to 76 - 85 mg · kg^{-1} in the saponites. Higher weathering levels in the C, Bw, and A horizons produce different Zn distributions depending on whether the control or the amended soil is considered.

When taking into account the mean standard errors, the Zn concentrations in saponites and montmorillonites appear almost constant throughout the control soil profile (Figure 4(b)). The behaviour of Zn in the amended soil appears much more complicated. Initially, the Zn concentrations measured in the saponites derived from amphiboles are very similar in the C horizons of the amended and control soil profiles: 72 to 82 mg · kg^{-1} in AS7, AS6, AS5 versus 72 to 81 mg · kg^{-1} in CS6 and CS5 (Table 4, Figure 4(b)). In a second stage, the Zn concentrations measured in the montmorillonites from the Bw and A horizons of the amended soil differ strongly from that in the control soil. The concentrations increase in the amended soil from 105 to 220 mg · kg^{-1} in AS4 and AS1 versus 68 to 82 mg · kg^{-1} in CS4 and CS1 (Table 4, Figure 4(b)).

Table 4.Mean EPMA of Zn (n analyses, mg · kg^{-1}) in clays from amphiboles and plagioclases weathering,and fissure infillings in control (CS) and amended (AS) soils.

Amphiboles	Saponites								Montmorillonites								Saponites						Montmorillonites							
Sample	CS8		CS7		CS6		CS5		CS4		CS3		CS2		CS1		AS7		AS6		AS5		AS4		AS3		AS2		AS1	
Depth (cm)	485		450		180		160		130		65		40		20		190		170		140		110		70		40		20	
	n = 4[a]	SE[b]	*n* = 4	SE	*n* = 4	SE	*n* = 4	SE	*n* = 4	SE	*n* = 4	SE	*n* = 4	SE	*n* = 4	SE	*n* = 4	SE	*n* = 4	SE	*n* = 4	SE	*n* = 4	SE	*n* = 4	SE	*n* = 4	SE	*n* = 4	SE
Zn	76	7	85	7	72	4	81	5	77	4	68	4	82	6	75	4	72	8	75	7	82	2	105	4	165	4	192	8	220	6
Plagioclases	Montmorillonites								K/S mixed layers				Kaolinites				Montmorillonites						K/S mixed layers				Kaolinites			
Sample	CS8		CS7		CS6		CS5		CS4		CS3		CS2		CS1		AS7		AS6		AS5		AS4		AS3		AS2		AS1	
Depth (cm)	485		450		180		160		130		65		40		20		190		170		140		110		70		40		20	
	n = 4	SE	*n* = 4	SE	*n* = 4	SE	*n* = 4	SE	*n* = 4	SE	*n* = 4	SE	*n* = 4	SE	*n* = 4	SE	*n* = 4	SE	*n* = 4	SE	*n* = 4	SE	*n* = 4	SE	*n* = 4	SE	*n* = 4	SE	*n* = 4	SE
Zn	77	15	66	10	42	4	39	5	37	4	35	3	31	3	27	2	67	8	64	7	39	3	36	4	32	4	25	2	30	2
Fissures	Montmorillonites								Montmorillonites + Kaolinites								Montmorillonites						Montmorillonites + kaolinites							
Sample	CS8		CS7		CS6		CS5		CS4		CS3		CS2		CS1		AS7		AS6		AS5		AS4		AS3		AS2		AS1	
Depth (cm)	485		450		180		160		130		65		40		20		190		170		140		110		70		40		20	
	n = 4	SE	*n* = 4	SE	*n* = 4	SE	*n* = 4	SE	*n* = 4	SE	*n* = 4	SE	*n* = 4	SE	*n* = 4	SE	*n* = 4	SE	*n* = 4	SE	*n* = 4	SE	*n* = 4	SE	*n* = 4	SE	*n* = 4	SE	*n* = 4	SE
Zn	156	8	172	9	165	8	138	7	115	6	58	3	64	4	45	3	268	10	254	6	185	6	156	8	162	7	246	9	280	11

[a]Each analysis is already the mean of ten measurements. [b]Standard error.

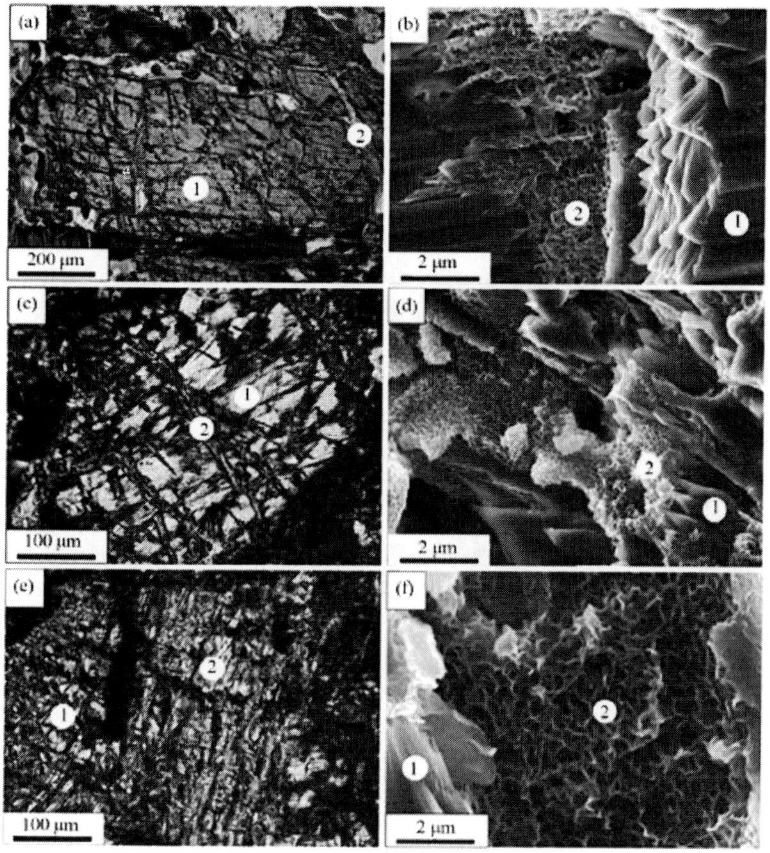

Figure 5. Optical Microscopy (OM) and Scanning Electron Microscopy (SEM) of the amphibole weathering. (a) OM view of an early weathering microsite in the C horizon of the control soil, with 1: elongated amphibole crystal; 2: intramineralmicrocrack with clay minerals infillings. (b) SEM view of the microsite a) with 1: amphibole crystal; 2: honeycomb texture of clay minerals in a microcrack. (c) OM view of a weathering microsite in the Bw horizon of the control soil with 1: amphibole fragments; 2: dense microcrack network with clay minerals infillings. (d) SEM view of the microsite c) with 1: saw-tooth amphibole terminations; 2: honeycomb clay minerals cover. (e) OM view of a weathering microsite in the A horizon of the control soil with 1: increasing amphibole fragmentation with clay minerals crystallizations; 2: complete replacement of amphibole fragments by a clayey plasma. (f) SEM view of the microsite e) with 1: amphibole fragment; 2: honeycomb-textured clay mineral.

3.5. Behaviour of Zn in the Plagioclase Weathering Microsystems

The first weathered plagioclases were observed under OM in the R and C horizons of the two soils. The heart of the plagioclases is dotted with etch pits which are filled by newly-formed clay minerals (Figure 6(a)). The SEM gives more details about these clay minerals which display a typical honeycomb habit, previously observed for the smectites in the amphiboles (Figure 6(b)). The plagioclases observed under OM in the Bw horizons show more advanced weathering stage. The etch pits, more numerous and connected, lead to the fragmentation of the plagioclase crystals into small pieces, 200 μm as largest size, with clay minerals patches (Figure 6(c)). The SEM reveals that these clay minerals have a spheroidal habit (Figure 6(d)) very different from the honeycombs observed in the first weathering stage. The most weathered plagioclases are observed in the A horizons. The plagioclase crystals are now almost entirely replaced by a light-grey clayey weathering plasma (Figure 6(e)). The SEM reveals that these clay minerals have an exfoliated "booklet-type" morphology (Figure 6(f)) which is commonly observed for the kaolinite crystallizations in the altered rocks [26] [27] .

The XRD patterns of the clay minerals extracted from the plagioclase grains differ according to their location in the two soil profiles. The clay minerals in the C horizon of the two soils are similar and made up of a smectite with a 15.56 Å reflection in Ca-oriented, air-dried state, shifting to 17.06 Å after ethylene-glycol solvation, with a second order reflection at 8.48 Å. The XRD patterns of the clay minerals with spheroidal habit (Figure 6(d)) in the Bw horizons are more complex, but similar in the two soils. At the bottom of the Bw horizons (140 - 130 cm depth) XRD characterizes a random kaolinite/smectite (K/S) mixed layer with 65% smectite component. This clay mineral is typified by its reflections at 15.03 Å, 7.36 Å, and 3.55 Å in Ca-oriented, air-dried state, shifting to 17.14 Å, 8.28 Å, and 3.42 Å after ethylene-glycol solvation. This K/S is replaced at the top of the Bw horizons (70 - 65 cm depth) by a second random K/S with only 5% smectite component. This K/S differs from the first one by its very low expansion at 7.36 Å (17.14 Å for the first K/S) in Ca-oriented, ethylene-glycol solvated state. The XRD of the clay minerals with the exfoliated "booklet-type" observed in the A horizons characterize a kaolinite clay mineral with a typical 7.17 Å-4.44 Å-3.57 Å reflection group in Ca-oriented, air-dried state, unchanged in the Ca-ethylene-glycol-solvated state.

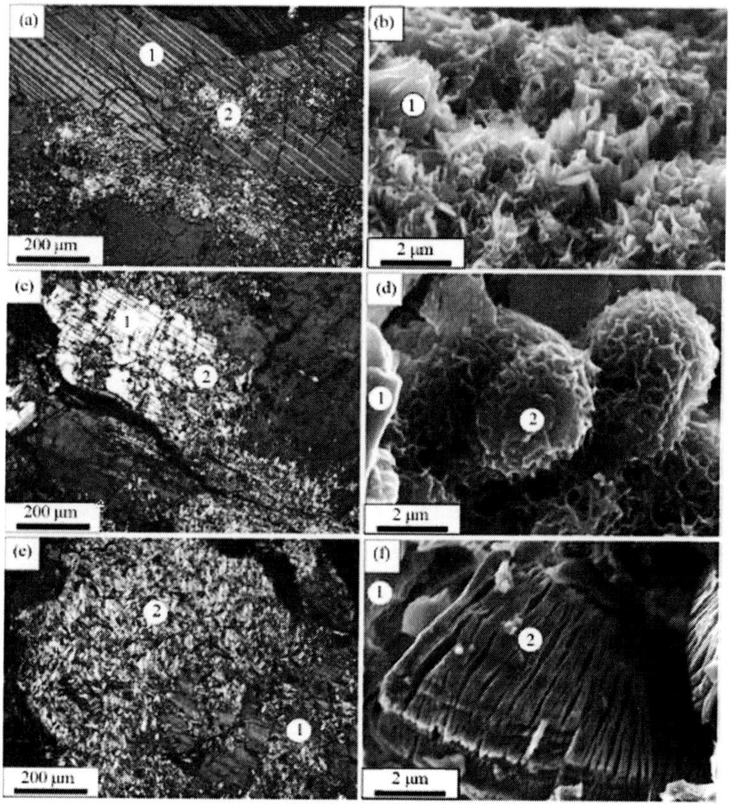

Figure 6. Optical Microscopy (OM) and Scanning Electron Microscopy (SEM) of the plagioclase weathering. (a) OM view of an early weathering microsite in the C horizon of the control soil, with 1: tabular plagioclase crystal; 2: clay minerals infilling dissolution pits. (b) SEM view of the microsite a) with 1: honeycomb texture of clay minerals in a dissolution pit. (c) OM view of a weathering microsite in the Bw horizon of the control soil with 1: plagioclase fragmentation; 2: clay minerals infillings in enlarged dissolution pits. (d) SEM view of the microsite c) with 1: plagioclase remnant; 2: spheroidal clay minerals filling a dissolution void. (e) OM view of a weathering microsite in the A horizon of the control soil with 1: increasing plagioclase fragme- ntation with clay minerals crystallizations; 2: complete replacement of plagioclase fragments by a clayey plasma. (f) SEM view of the microsite e) with 1: plagioclase fragment; 2: "booklet-type"-textured clay mineral.

The EPMA of the clay minerals in the weathered plagioclases are in good agreement with the XRD results. The chemical analyses characterize two weathering sequences producing different clay minerals. The first sequence appears in the parent-rock and the C horizons of soils with the early weathering of plagioclases into aluminous smectites of the montmorillonite group. The second sequence occurs in the Bw and A horizons where montmorillonite weathers first into K/S (Bw horizons) and finally into kaolinite (A horizons).

The Zn concentrations in the clay minerals produced by the weathering of plagioclases are given inTable 4 and plotted versus depth in Figure 4(c). The early plagioclase weathering into montmorillonites leads to a general Zn enrichment from 17 mg · kg^{-1} in the fresh plagioclase (Table 2) to 77 - 66 mg · kg^{-1} (R horizons, samples CS8 and CS7). The Zn distribution in the C, Bw, and A horizons show similar trends in the control and amended soils. The Zn concentrations in the control soil decrease regularly from 42 mg · kg^{-1} at 180 cm depth to 27 mg · kg^{-1} at the soil surface (20 cm depth). A similar evolution is observed in the amended soil, but concentrations in the deepest levels are higher: 67 mg · kg^{-1} at 190 cm and 64 mg · kg^{-1} at 170 cm depths (Figure 4(c)). In both soils, the decrease in Zn concentrations from bottom to top follow the decrease in smectite contents observed in the clay minerals: 100% smectite in the C horizons (montmorillonites), followed by 65% and 5% smectite in the K/S from the Bw horizons and 0% smectite in the kaolinite from the A horizons.

3.6. Behaviour of Zn in the Fissural Microsystems

The fissural microsystems are made up of grey to light-green clay minerals that aggregate in the fissural network of the weathered rock and soil. This fissural network ranges from 10 μm to millimeter in width and the clay minerals are observed as infilling clay material. Since clay minerals could not be separated from the bulk soil samples in sufficient pure amount to use XRD identification, they were characterized in situ from thin sections using EPMA. The clay mineralogy deduced from the EPMA of the major elements evolves from pure montmorillonite at the bottom of the two soils to a mixture of montmorillonite and kaolinite at the top. The vertical distribution profile of Zn (Figure 4(d)) obtained from the EPMA of the clay minerals (Table 4) shows firstly the anthropogenic contamination of the fissural clays by Zn in the amended soil surface, when compared to the control soil. The Zn concentrations measured in the amended soil are always higher than in control soil (156 to 280 $mg \cdot kg^{-1}$ versus 45 to 172 $mg \cdot kg^{-1}$, Table 4). The vertical distribution of Zn in the control soil is typical of a downward migrating element, in good correlation with the clay mineralogy (Figure 4(d)). The lowest content (45 $mg \cdot kg^{-1}$, Table 4) is measured in the fissural clays from the surface sample (sample CS1, 20 cm depth) which are dominated by the kaolinite with very low sorption capacity. Zn is then allowed to migrate downwards, to be retained in the level where fissural infillings are montmorillonites with high sorption capacities. This occurs from 160 cm depth in the fissure of the sample CS5 (138 $mg \cdot kg^{-1}$) down to 485 cm in the sample CS8 (156 $mg \cdot kg^{-1}$). The vertical distribution of Zn in the amended soil is very similar but with higher concentrations and Zn accumulation in the surface horizons, i.e. 280 - 246 $mg \cdot kg^{-1}$ in samples AS1 and AS2 at 20 - 40 cm depth (Table 4). This surface accumulation, not observed in the control soil, can be directly related to the spreading of the sewage sludges.

4. DISCUSSION AND CONCLUSIONS

In the first place, two meaningful observations must be discussed at the scale of the soil profiles. The first observation is that the control and the amended soils have the same diorite parent-rock and developed C, Bw and A horizons with similar physiochemical properties, unmodified by sludge spreading (Figire 2(d)). This is particularly true for the pH which can affect the variable-charge sites in the clay minerals through protonation or deprotonation of the hydroxyl (SOH) groups. The higher the pH, the higher the adsorption of the heavy metals at the clay-water interfaces [28] -[31] and the lower their mobility [32] . The pH values measured in the amended soil are even closer to neutrality (6.5 - 6.8) than in the control soil (6.1 - 6.5), and thus it will favor adsorption of the heavy metals at the clay surfaces. The second observation is that the geochemical mass balances display the same major elements distribution in the two soil profiles (Figure 3(a), Figure 3(b)). This distribution is typical of similar weathering and pedogenetic processes with increasing LOI content (crystallization of newly-formed clay minerals), alkaline earth leaching (weathering of primary rock-forming minerals) and stability of SiO_2 and Fe_2O_3. This latter observation indicates that SiO_2 and Fe_2O_3, initially released by the weathering of the rock- forming minerals, are precipitated in the newly-formed clay minerals and do not leave the soil profiles.

These observations show that the two soils are formed from the same parent-rock and have suffered similar weathering and pedogenetic processes which lead to the same horizons development and the same weathering clay mineralogy (smectites, K/S, kaolinite). These converging data validate the choice of these two soils to follow the behaviour of the Zn when it is applied to the amended soil by long-term sludge spreading.

The bulk Zn concentrations in the soil samples (Table 1) remain always below the value of 300 $mg \cdot kg^{-1}$, upper threshold value for non-polluted soil in the French regulation. Thus, repeated sludge spreadings on the area do not induce Zn pollution. The distribution of the EF-Zn versus depth in the amended soil (Figure 3(c)) shows a pronounced Zn depletion from the surface horizons which weakens downward and indicates Zn migration down to 190 cm depth. These data reflect the behaviour of zinc at the scale of the soil sample, but a further study of the soil components indicates that local Zn concentrations may occur in specific soil compartments such as the <2 μm clay fraction, the weathering microsites at the scale of the rock-forming minerals, or the fissural network.

The Zn concentrations in the <2 μm clay fractions of the amended soil (Table 3) reveal clearly the anthropogenic contamination of the clay minerals in the surface horizons (20 to 40 cm depth) when compared to the control soil. This surface contamination is associated with the downward Zn migration down to 190 cm depth (Figure 4(a)). These clay fractions, however, are the collection of the clay minerals which have formed during the weathering of the rock-forming minerals or have migrated in the fissural network. Thus, the study of Zn behaviour in the two soils will be greatly improved by considering the clay minerals in their crystallization sites, i.e. at the micrometric scale of the thin section.

The Zn distribution profiles collected from the amphiboles weathering microsites (Figure 4(b)) reveal a general anthropogenic contamination of the amended surface horizons with Zn. This contamination can be related to the crystallization of large amount of montmorillonite as weathering product of amphibole. This montmorillonite, with high sorption capacity of 90 to 127 cmol · kg^{-1}[33] [34] may fix a part of Zn provided by the sludges in the surface horizons. This surface accumulation is followed by a regular Zn decrease in the deeper horizons of the amended soil, in relation with the decrease in amphibole weathering intensity and, accordingly, a decrease in smectite content and sorption capacity with depth.

Conversely, the Zn concentration profiles obtained from the plagioclases weathering microsites (Figure 4(c)) show a Zn depletion of the clay minerals in the amended surface horizons and a Zn enrichment deeper in the amended soil. This particular behaviour appears to be controlled by the clay mineral weathering sequence of plagioclases. Kaolinite with low sorption capacity of 16 to 34 cmol · kg^{-1} [35] [36] crystallizes at the top of the amended soil and does not retain Zn. Thus, Zn can migrate downwards to more smectitic assemblages (K/S) and finally to montmorillonite where it is retained at 190 cm depth.

Although Zn concentrations in the fissural network of the amended soil are always higher than in the control soil, the general trend is similar for the C and Bw horizons (Figure 4(d)) and typical of a downward migrating element. The behaviour of Zn in the surface horizons appears very different depending on whether one considers the control or the amended soil: Zn in the control soil behaves like a migrating element and follows the same trend as in C and Bw horizons; conversely, Zn in the amended soil shows a surface accumulation that can be directly related to the spreading of the sewage sludges.

Regardless of the sites studied, the present work points out the migration of Zn within the sludge-amended soil profile. The mineralogical multiscale approach demonstrates that such a behaviour is highly controlled by the clay minerals and their associated sorption capacities. Clays study is thus essential to understand the sorption reactions occurring at the solid/solution interface in soils. As demonstrated in this paper, the clayey components in the soils are complex assemblages of several clay minerals which formed in distinct weathering microsystems with their own physicochemical properties. Therefore, there is a need for accurate studies at the scale of these weathering microsystems to identify each clay mineral species in order to infer the sorption capacity of the weathering microsite, and to estimate the potential mobility of the heavy metals in the contaminated soils. This micrometric mineralogical approach could be relevant to describe the location and the behaviour of heavy metals in contaminated soils and improve the modeling of the heavy metals surface complexation in soils [37] .

REFERENCES

1. Planquart, P., Bonin, G., Prone, A. and Massiani, C. (1999) Distribution, Movement and Plant Availability of Trace Metals in Soils Amended with Sewage Sludge Compost: Application to Low Metal Loadings. Science of the Total Environment, 241, 161-179.

2. Baize, D. and Sterckeman, T. (2001) Of the Necessity of Knowledge of the Natural Pedo-Geochemical Background Content in the Evaluation of the Contamination of Soils by Trace Elements. Science of the Total Environment, 264, 127-139.

3. Horckmans, L., Swennen, R., Deckers, J. and Maquil, R. (2005) Local Background Concentrations of Trace Elements in Soils: A Case Study in the Grand Duchy of Luxembourg. Catena, 59, 279-304.

4. Sterckeman, T., Douay, F., Baize, D., Fourier, H., Proix, N. and Schvartz, C. (2006) Trace Elements in Soils Developed in Sedimentary Materials from Northern France. Geoderma, 136, 912-929.

5. Huang, B., Kuo, S. and Bembenek, R. (2004) Availability of Cadmium in Some Phosphorous Fertilizer s to Field-Grown Lettuce. Water, Air, and Soil Pollution, 158, 37-51.

6. Mbila, M.O., Thompson, M.L., Mbagwu, J.S.C. and Laird, D.A. (2001) Distribution and Movement of Sludge-Derived Trace Metals in Selected Nigerian Soils. Journal of Environmental Quality, 30, 1667-1674.

7. Hernandez, L., Probst, A., Probst, J.L. and Ulrich, E. (2003) Heavy Metal Distribution in Some French Forest Soil: Evidence for Atmospheric Contamination. Science of the Total Environment, 312, 195-219.

8. Kraepiel, A.M.L., Keller, K. and Morel, F.M.M. (1999) A Model for Metal Adsorption on Montmorillonite. Journal of Colloid and Interface Science, 210, 43-54.

9. Srivastava, P., Singh, B. and Angove, M. (2005) Competitive Adsorption Behavior of Heavy Metals on Kaolinite. Journal of Colloid and Interface Science, 290, 28-38.

10. Zhu, J., Cozzolino, V., Pigna, M., Huang, Q., Caporale, A.G. and Violante, A. (2011) Sorption of Cu, Pb and Cr on Na-Montmorillonite: Competition and Effect of Major Elements. Chemosphere, 84, 484–489.

11. Farrah, H., Hatton, D. and Pickering, W.F. (1980) The Affinity of Metal Ions for Clay Surfaces. Chemical Geology, 28, 55-68.

12. Wahba, M.M. and Zaghloul, A.M. (2007) Adsorption Characteristics of Some Heavy Metals by Some Soil Minerals. Journal of Applied Sciences Research, 3, 421-426.

13. Wilson, M.J. (2004) Weathering of the Primary Rock-Forming Minerals: Processes, Products and Rates. Clay Minerals, 39, 233-266.

14. Caillaud, J., Proust, D. and Righi, D. (2006) Weathering Sequences of Rock-Forming Minerals in a Serpentinite: Influence of Microsystems on Clay Mineralogy. Clays and Clay Minerals, 54, 87-100.

15. Proust, D., Caillaud, J. and Fontaine, C. (2006) Clay Minerals in Early Amphibole Weathering: Trito Dioctahedral Sequence as a

Function of Crystallization Sites in the Amphibole. Clays and Clay Minerals, 54, 351-362.

16. Keller, C., McGrath, S.P. and Dunham, S.J. (2002) Trace Metal Leaching through a Soil-Grassland System after Sewage Sludge Application. Journal of Environmental Quality, 31, 1550-1560.

17. Martinez Cortizas, A., Garcia-Rodeja Gayoso, E., Novoa Munoz, J.C., Pontevedra Pombal, X., Burman, P. and Terribile, F. (2003) Distribution of Some Selected Major and Trace Elements in Four Italian Soils Developed from the Deposits of the Gauro and Vico Volcanoes. Geoderma, 117, 215-224.

18. Atteia, O., Thélin, P., Pfeifer, H.R., Dubois, J.P. and Hunziker, J.C. (1995) A Search for the Origin of Cadmium in the Soil of the Swiss Jura. Geoderma, 68, 149-172.

19. Hardy, M. and Cornu, S. (2006) Location of Natural Trace Elements in Silty Soils Using Particle-Size Fractionation. Geoderma, 133, 295-308.

20. Camuti, K.S. and McGuire, P.T. (1999) Preparation of Polished Thin Sections from Poorly Consolidated Regolith and Sediment Materials. Sedimentary Geology, 128, 171-178.

21. Fialin, M., Rémy, H., Richard, C. and Wagner, C. (1999) Trace Element Analysis with the Electron Microprobe: New Data and Perspectives. American Mineralogist, 84, 70-77.

22. Baize, D. (1997) Teneurs totales en Eléments Traces métalliques dans les sols (France). INRA, Paris, 408 p.

23. Scokart, P.O., Meeus-Verdinne, K. and De Borger, R. (1983) Mobility of Heavy Metals in Polluted Soils. Water, Air, and Soil Pollution, 20, 451-463.

24. Arslan, M., Abdioglu, E. and Kadi, S. (2010) Mineralogy, Geochemistry, and Origin of Bentonite in Upper Cretaceous Pyroclastic Units of the Tirebolu Area, Giresun, Northeast Turkey. Clays and Clay Minerals, 58, 120-141.

25. Karakaya, M.C., Karakaya, N. and Kupeli, S. (2011) Mineralogical and Geochemical Properties of the Na- and Ca-Bentonites of Ordu (NE Turkey). Clays and Clay Minerals, 59, 75-94.

26. Jeong, G.Y. (2000) The Dependence of Localized Crystallization of Halloysite and Kaolinite on Primary Minerals in the Weathering Profile of Granite. Clays and Clay Minerals, 48, 196-203.

27. Papoulis, D., Tsolis-Katagas, P. and Katagas, C. (2004) Progressive Stages in the Formation of Kaolin Minerals of Different Morphologies in the Weathering of Plagioclase. Clays and Clay Minerals, 52, 275-286.

28. Coles, C.A. and Yong, R.N. (2002) Aspects of Kaolinite Characterization and Retention of Pb and Cd. Applied Clay Science, 22, 39-45.

29. Eloussaief, M. and Benzina, M. (2010) Efficiency of Natural and Acid-Activated Clays in the Removal of Pb(II) from Aqueous Solutions. Journal of Hazardous Materials, 178, 753-757.

30. Farrah, H. and Pickering, W.F. (1977) Influence of Clay-Solute Interactions on Aqueous Heavy Metal Ion Levels. Water, Air, and Soil Pollution, 8, 189-197.

31. Veli, S. and Alyüz, B. (2007) Adsorption of Copper and Zinc from Aqueous Solutions by Using Natural Clay. Journal of Hazardous Materials, 149, 226-233.

32. Kirkham, M.B. (2006) Cadmium in Plants on Polluted Soils: Effects of Soil Factors, Hyperaccumulation, and Amendments. Geoderma, 137, 19-32.

33. Ayari, F., Srasra, E. and Trabelsi-Ayadi, M. (2005) Characterization of Bentonitic Clays and Their Use as Adsorbent. Desalination, 185, 391-397.

34. Srodon, J. and McCarty, D.K. (2008) Surface Area and Layer Charge of Smectite from CEC and EGME/H2O-Retention Measurements. Clays and Clay Minerals, 56, 155-174.

35. Ferris, A.P. and Jepson, W.B. (1975) The Exchange Capacities of Kaolinite and the Preparation of Homoionic Clays. Journal of Colloid and Interface Science, 51, 245-259.

36. Ma, C. and Eggleton, R.A. (1999) Cation Exchange Capacity of Kaolinite. Clays and Clay Minerals, 47, 174-180.

37. Hizal, J. and Apak, R. (2006) Modeling of Copper (II) and Lead (II) Adsorption on Kaolinite-Based Clay Minerals Individually and in the Presence of Humic Acid. Journal of Colloid and Interface Science, 295, 1-13.

CHAPTER 4

Effects of Land Use Change on Land Degradation Reflected by Soil Properties along Mara River, Kenya and Tanzania

Ally-Said Matano[1,2], Canisius K. Kanangire[1], Douglas N. Anyona[2], Paul O. Abuom[2], Frank B. Gelder[3], Gabriel O. Dida[4*], Philip O. Owuor[5], Ayub V. O. Ofulla[4]

[1]Lake Victoria Basin Commission Secretariat, Kisumu, Kenya

[2]School of Environment and Earth Sciences, Maseno University, Maseno, Kenya

[3]Probe International, Inc., Auckland, New Zealand

[4]School of Public Health and Community Development, Maseno University, Maseno, Kenya

[5]Department of Chemistry, Maseno University, Maseno, Kenya

ABSTRACT

Human-induced changes to natural landscapes have been identified as some of the greatest threats to freshwater resources. The change from natural forest cover to agricultural and pastoral activities is rampant especially in the upper Mara River catchment (water tower), as well as along the course of the Mara River. The objective of this study was to determine the effect of land use change on the physico-chemical properties of soil (bulk density, carbon, nitrogen, phosphorus and pH) along the course of the Mara River. Five major land uses (agricultural lands, livestock/pastoral lands, forested lands, conservancy/game reserves, and natural wetland) were explored. Results revealed that the mean soil bulk density was 0.956 g/cm^3 and differed significantly between sites ($p < 0.001$). Live biomass values differed significantly between sampling sites (land use types) within the Mara River Basin ($F_{(4, 147)} = 8.57$, $p < 0.001$). The mean infiltration over a period of 150 minutes differed, not only among sampling sites, but also between different sides of the river (left and right) within the same sampling site. Soil pH was generally acidic across the five sites and varied significantly ($F_{(4, 63)} = 19.26$, $p < 0.0001$) between sites along the Mara River Basin. The mean percentage soil nitrogen across all sampling blocks was 4.87%, with significant differences observed in percentage soil nitrogen ($F_{(4, 63)} = 3.26$, $p < 0.006$) between sampling sites. The results indicated that the five land use types affected land degradation differently along the Mara River, while adjacent land degradation affected water physico-chemical properties. These results point to the need to have focused policies on integrated land and water resource management strategies in the Mara River Basin.

KEYWORDS

Catchment Area, land Degradation, Land Use Types, Mara River, Riparian Land

1. INTRODUCTION

Land use and land cover changes associated with human activities and natural factors compromise many ecosystems including watersheds of important rivers [1] . Land degradation resulting from human activities has been a major global challenge since the 20th century and will remain high on the international agenda in the 21st century [2] . According to

Bai et al. [3] , land degradation is increasing in severity and extent in many parts of the world, with more than 20% of all cultivated areas, 30% of forests and 10% of grasslands undergoing degradation. In recent centuries, an increasing amount of riparian lands have been developed and utilized for agriculture, human settlements and development of cities and towns [4] . This has significantly impacted on critical catchment areas, thus altering water quality in aquatic ecosystems.

Land degradation encompasses the whole environment including individual factors such as soils, water resources (surface and ground), forests (woodlands), grasslands (rangelands), croplands (rainfed and irrigated) and biodiversity (animals, vegetative cover and soil) [5] . Different studies have examined the effects of land use/ cover change on soil physico-chemical properties, and most concur that despite its consequences vary, land use change frequently leads to nutrient losses and reduction of organic matter inputs in the soil [6] -[8] . Conversion of natural forest to other forms of land uses such as farmlands and pasturelands can provoke soil erosion and lead to a reduction in soil nutrients and modification of soil structure [9] . Rai and Sharma [10] also concur that change in land use types negatively affects soil productivity characteristics such as soil bulk density and hydraulic conductivity. Cultivation of forests for instance can diminish soil carbon (C) within a few years of initial conversion [11] and substantially lower mineralizable nitrogen (N) [12] . Islam and Weil [13] reported an increase in bulk density and a reduction in porosity and aggregate stability following the conversion of forest land to crop land, with consequent degradation of adjacent aquatic system.

Soil erosion, salinity and absence of vegetation cover are early warning signs of land degradation, which are likely to influence adjacent aquatic systems through sediment loading. The relationship between land use and water quality has also been demonstrated over the last two decades [14] [15] . There is convincing evidence that watersheds dominated by agriculture and/or human settlement have significantly higher river and nutrient levels [15] [16] . Since rivers provide many ecological and social service functions, they are subject to increased human exploitation and pollution [17] . A river system degrades under severe interference by anthropogenic activities in the catchment [4] . Interference can either occur directly on the river itself or indirectly by degradation of adjacent riparian land which then impacts on the aquatic system. Numerous studies have demonstrated the effects of land use change on erosion and sediment loading patterns in aquatic ecosystems (Alin et al. [18]). The spatial relationship between land use and water quality has also been examined by many researchers (Tong and Chen [19] , Ngoye and Machiwa [20] and King et al. [21]). More specifically, some investigators have compared land cover within certain distances from a stream or sampling site [22] . Burcher et al. [1] also observed that changes in land use and land cover interact with anthropogenic and natural drivers to impact negatively on the water quality of watersheds.

Three important spatial scales that influence physical, chemical and biological conditions in a river are basin-wide conditions, riparian (area adjacent to the stream) conditions, and in-stream conditions. The impact of land use change on the watershed environment has also been reported to vary across different spatial scales [23] . Land use changes therefore have both direct and indirect effects on freshwater ecosystems with the former having immediate ecological impacts (e.g. destruction of wildlife habitats), while the latter has impacts that are normally transmitted via altered flow or sediment transport patterns (e.g. lower productivity due to increasing turbidity). Sediment deposition in adjacent water bodies is driven by soil erosion, which is the most widely recognized and most common form of land degradation. Overgrazing of rangelands, poor cultivation of croplands, deforestation and urbanization are some of the land use practices that result in increased soil erosion and subsequent load of sediments and nutrients into aquatic systems [24] .

Land use changes in the Lake Victoria Basin have transformed land cover to mainly farmlands, grazing lands, human settlements and urban centres from the previous natural vegetation cover [9] . Over the last 50 years, the Mara River basin has undergone major changes in land use and land cover. Forests and savannah grasslands have been cleared and turned into land with the main purpose of expanding agricultural activities [25] . Many researchers, including Mugisha [26] , Misana et al. [27] and Olson et al. [28] , all concur that most of the changes observed in land use/cover in many parts of Africa are mainly associated with extension and intensification of agricultural activities to new areas. Livestock and wildlife grazing has also been cited as a source of soil degradation. Over cultivation and overgrazing have been linked to increased nutrient transfer and bulk density through nutrient loss and compaction of the soil leading to accelerated soil erosion [29] . High soil bulk density has been a serious land degradation problem in the entire Lake Victoria basin landscape including the Mara River Basin, due to unsustainable land uses [30] . Over the last 50 years, the Mara River Basin has undergone unprecedented changes in land use, just like many other river basins within the larger Lake Victoria Basin (LVB). Accelerated loss of vegetation at the upper Mara River basin has been reported in several studies, including Mati et al. [31] and WWF [32] .

The Mara River Basin (MRB) plays a major ecological and socio-economic role in communities living in the basin. There is growing evidence of land degradation in the MRB due to improper land use practices, which directly impacts on adjacent aquatic ecosystems. According to Dwasi [25] , forests and savannah grasslands in MRB have been cleared and turned into agricultural lands. Rapidly increasing population in the Mara River Basin puts an even greater pressure on the limited natural resources, resulting in increased pollution. Records from the Government of Kenya [33] indicate that over 7000 hectares of Mau forest, which is one of the major water towers in Kenya, were destroyed between 2000 and 2003. The records further show that the area under cultivation in the Amala sub catchment increased from less than 20% in 1960 to more than 51% in 1991 to give way to Olenguruone Settlement Scheme. Such changes in land use result in increased degradation of water quality, thus affecting aquatic biota (flora and fauna). The Lake Victoria Basin region has witnessed increased land use changes in recent times, some of which have led to accelerated land degradation reflected through diminishing vegetation cover, reduced biomass, increased bulk density, and reduced soil nutrients among others [31] [33] . A recent study by USAID EA [34] also linked land use practices in the Mara River Basin to environmental flows and water quality degradation along the Mara River. Over time, the Mara River Basin has witnessed intensified land use changes resulting in increased pollution of the Mara River waters, visible through the high turbidity, total dissolved and total suspended solids, which is an indication of off-site effects of soil erosion. This has resulted in changes in the physical, chemical and biological properties of water in the Mara River ecosystem. However, the exact link between land use types and soil properties along the Mara River Basin has not been well established. This study therefore set out to determine the effects of land use types on land degradation as reflected by soil properties (soil bulk density, soil organic carbon, nitrogen, and phosphorus) along the Mara River.

2. MATERIALS AND METHODS

2.1. Study Area Description

The Mara River basin covers 13,750 km^2 and lies between Kenya and Tanzania. The Basin is located roughly between longitudes 33°47'E and 35°47'E and latitudes 0°38'S and 1°52'S, with the upper 65% area (8941 km^2) being in Kenya and the remaining 35% in Tanzania (Figure 1). The Mara River originates from the Napuiyapui swamp in the Mau escarpment in the highlands of Kenya, with altitudes ranging from 2932 m at its source to 1134 m at Musoma bay.

Rainfall varies inter-annually by a factor of about four between extreme wet and dry years [35] . The river which has for a long time been considered as one of the more pristine rivers draining into Lake Victoria [36] , traverses through different land use types including forests, farmlands, open lands, urban centers, game reserves and conservancy before flowing through the Mara Swamp at Musoma Bay in the lower Mara and finally into Lake Victoria. Five distinct land use types along the Mara River were selected for this study. These were: Silibwet sampling site (forested land but with some human interference), Kapkimolwa (agropastoralism with subsistence and large scale farming), Bomet town sampling site (urban set-up with high population and economic activities), Ngerende sampling site (protected area/conservancy) and Kirumi bridge sampling site (a natural wetland with relatively low human interference).

2.2. Study Design Based on Land Degradation Surveillance Framework (LDSF)

A modified version of the LDSF was used in this study. The cross sectional field component of this study was carried out between July 2011 and September 2011 during which soil samples were collected for analysis in established laboratories at the International Centre for Research in Agroforestry (ICRAF), Kisumu, Kenya and the Kenya Forestry Research Institute (KEFRI) in Maseno, Kenya. In the study, the entire Mara River Basin represented a block with 5 sampling sites, namely, Silibwet, Kapkimolwa, Bomet, Ngerende and Kirumi, and purposively selected to represent different predominant land uses. In each of the sampling sites there were 8 plots (four on each side of the river) which were laid on a line transect on either side of the river. Each sampling site was also projected to within a 5 km radius from a designated central point within the river, out of which 4 plots were selected within an area lying between 22.5° and 45° degrees on either side of the river. The left and right sides of the river were determined with the researcher facing downstream (Figure 2).

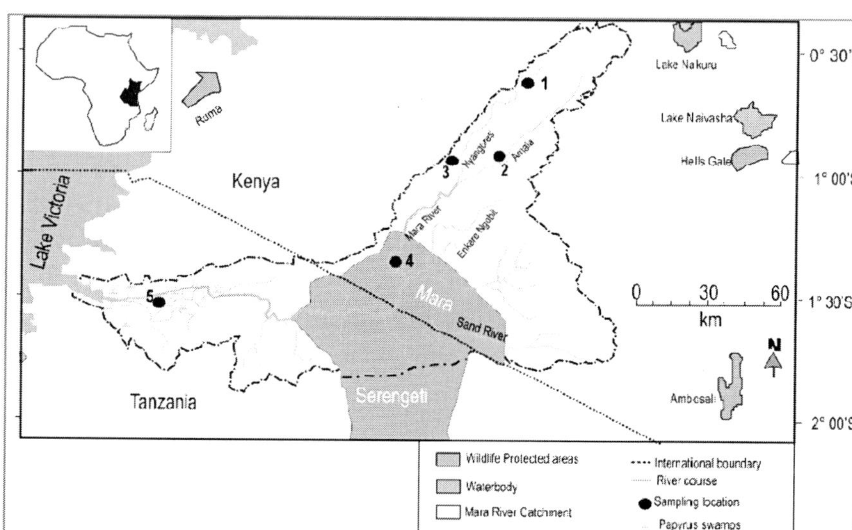

Figure 1. Map of the Mara River basin showing sampling blocks numbered 1 - 5.

Thereafter, each of these sites was laid out as a straight-line transect, with sampling distances for the 4 plots doubling from the central point up to a maximum of the 5 km limit. The same was repeated on the other side of the river. For instance a 3 km transect had the first plot (L1 or R1) 200 m from the central point, with the second (L2 or R2) being located 400 m from the first plot (600 m from the central point) and the third 800 m from the second plot (1.4 km from the central point). The final plot in such transect (L4 or R4) was 1.6 km from the third plot, and thus 3 km from the water sampling point. In all cases the sampling point was located at a bridge.

Infiltration rates were determined in situ at each sampling point, while soil samples were collected for determination of soil bulk density, soil carbon, % soil nitrogen, % soil phosphorus and soil pH in the laboratory. A total of 16 soil samples, 8 from the left and 8 from the right side of the river at each sampling site were collected making a total of 80 samples for soil pH, soil carbon, and percentage nitrogen and phosphorus determination. Modifications aside, all other LDSF protocols were observed.

2.3. Land Degradation Surveillance Framework Sampling Plan

In this study, plots measuring 1000 m^2, 4 subplots (SP) measuring 100 m^2 (Figure 3) and their equidistance were established from each of the 5 sampling sites. Each subplot yielded at least 1 top-soil (0 - 20 cm) and 1 sub-soil (20 - 50 cm) aggregate samples (yielding at least 8 soil sample replicates per plot). Bulk Density (BD) sample collection was done at SP 1 (centre) using a standard Bulk Density (BD) ring, and weight readings were determined on-site. In addition, soil hydraulic conductivity tests (infiltration) were carried out at SP 1 (center) in 8 plots within the sampling site thus providing data for 40 soil conductivity samples.

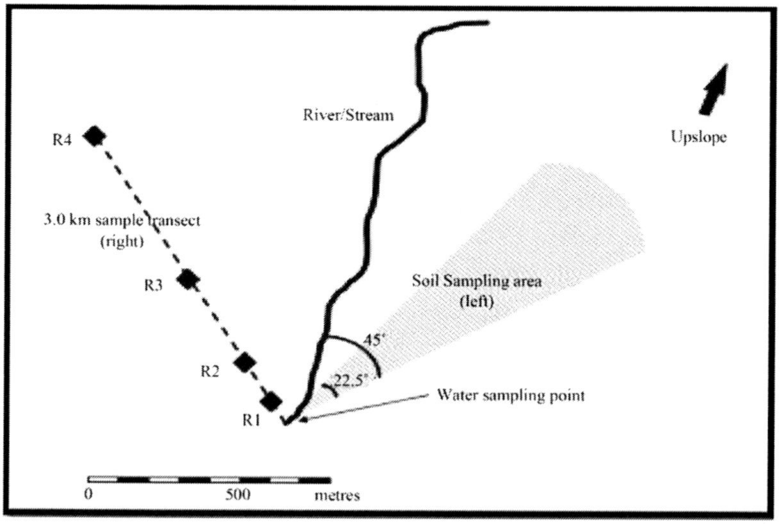

Figure 2. Sampling plot within a sampling site.

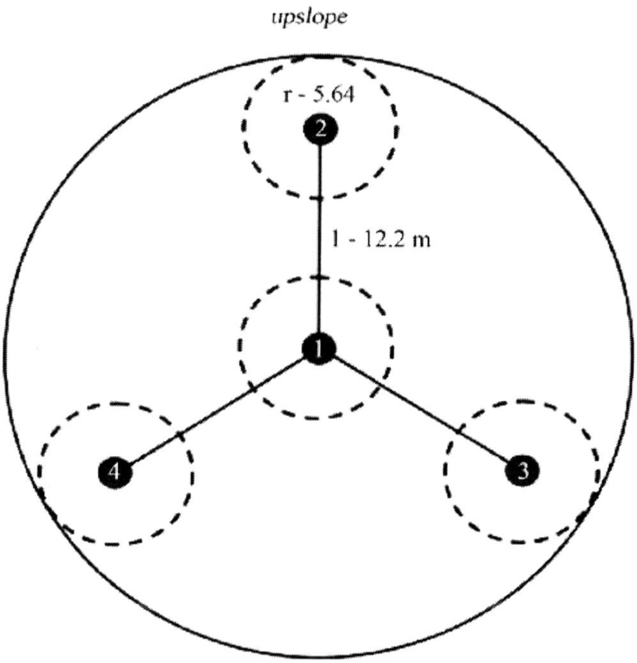

Figure 3. A 1000 m^2 LSDF plot layout showing the four 100 m^2 sub-plots and distances.

2.4. Soil Sampling for Land Degradation Analysis

Soil sampling was done on both sides of the river within the area lying between the bearings of 22° - 45° upstream from the water sampling site as depicted in Figure 2 (R1 to R4 and L1 to L4). Soil samples for the soil pH, soil organic carbon (SOC), soil nitrogen, soil phosphorus, and soil bulk density were collected from the river catchment sites using a Dutch auger. Care was taken to avoid mixing the top and sub soil samples, and also to avoid collapsing the hole during sampling. As such, augering was done by taking small and steady bites and by keeping the auger as vertical as possible. Once a significant amount of soil was collected, it was emptied into a suitably labeled bucket to avoid overfilling the auger. In each site, 4 plots of 1000 m^2 were identified systematically. From each plot four sub plots were identified from which top soil (0 - 20 cm) and subsoil (20 - 50 cm) were collected giving 8 different soil samples replicates (4 topsoil and 4 subsoil samples) packed in 8 different polythene bags sealed with rubber band for laboratory analysis at the Kenya Forestry Research Institute (KEFRI) laboratory at Maseno, Kenya. Each polythene bag was clearly labeled with catchment name, area, plot, position and depth.

2.5. Soil Sampling for Land Degradation Analysis

Soil samples for determining soil nitrogen and soil phosphorous were air dried by spreading the sample out as a thin layer into shallow trays. Maximum care was taken to ensure that no materials from the sample were lost or discarded. Drying time depended on the condition of samples and ambient conditions, but they were nevertheless dried thoroughly. The choice for air dried samples was preferred because of the convenience in handling as well as to reduce variation due to moisture [37] . The air dried samples were then ground gently with a wooden pestle and mortar and pressed through a 2 mm sieve [38] . In line with Ben-Dor and Banin [37] , the screened sample was subjected to reflectance spectrometer which provided non-destructive rapid prediction of soil physical, chemical and biological properties in the laboratory at KEFRI, Maseno, Kenya.

2.6. Determination of Soil Bulk Density

Bulk Density (BD) samples were collected at each of Sub-Plot 1 using a standard Bulk Density (BD) ring. This involved a cylindrical metal sampler (52 mm high and 51 mm diameter) being driven into the soil until at the same level with ground. The sample was removed by digging around the ring with the trowel underneath it to prevent any loss of soil. Excess soil from the sample was removed with a flat bladed knife and the bottom of the sample made flat and even with the edges of the ring.

The total sample was put into a polythene bag for laboratory analysis and sealed with rubber band. Each polythene bag was identified by the sample site identification code. Top soil bulk density was used to characterize differences in soil compaction among different land use types in the landscape. A total of 32 soil sample replicates were collected per cluster (1 top soil sample × 4 sub-plots × 4 plots × 2 sides of the river) during the field work. Coarse organic matters were removed manually from bulk density soil samples. In the laboratory at International Centre for Research in Agroforestry (ICRAF), Kisumu, Kenya, the samples were oven dried completely at 105°C, weighed on sensitive weighing machine, and weight recorded in grams. The bulk density was then calculated based on mass/volume ratio of the bulk density sampling ring and values recorded.

2.7. Determination of Soil Infiltration Rate

All infiltration measurements were carried out at the mid sub-plots (SP 1) along the sampling transects. Eight (8) measurements were made per sampling site (1 mid subplot × 4 plots × 2 sides of the river). Surface saturated hydraulic conductivity was measured using single-ring cylinder, 16 cm inner diameter according to Reynolds and Elrick [39] taking into account ring radius, depth of ring insertion and depth of ponding. The test calculations were based on shape factors that suggest that field saturated hydraulic conductivity could be obtained with accuracy of about ±20% [39] . The ring was driven into a pre-wetted soil surface according to procedure reported by Reynolds and Elrick [39] .

The ring was then filled with water up to a given initial depth and the initial water depth noted while a given time interval (in minutes) was set.

At each and every interval the water depth was noted and the ring refilled to the initial depth. Using a stopwatch, readings were taken first, at 5 minute intervals (where possible) for at least the first 30 minutes, then the interval were increased to 10 minutes (1 hour), and finally to 20 minutes (1 hour or until the readings had stabilized). After two and half to three hours of ponding water in the ring, steady infiltration (Q_s, in mm per hour) and steady ponding depth (W, in cm), the depth of ring insertion (d in cm), was recorded for the ring. The determination of field saturated hydraulic conductivity was based on mass balance equation of flow into soil. Using mass balance, the steady infiltration from a single ring (Q_s) in mm/hr was approximated by the formula: $Q_s = Q_p + Q_g$, where Q_p and Q_g are the steady water out flows from the ring due to hydrostatic and capillary pressure, and due to gravity, respectively.

3. STATISTICAL METHODS

The field data was first entered in to Excel spreadsheet and analyzed for descriptive statistics. Further analysis was done using various statistical packages, including Genstat and the Statistical Package for Social Sciences (SPSS) software version 11.0. One way analysis of variance (ANOVA) was performed to test for significant differences among different soil and water quality parameters in relation to the five different land use types along the Mara River. If the main effects were found to be significant at $p < 0.05$, a post hoc separation of means analysis was done by means of Duncan Multiple Range Test (DMRT) to further elucidate the specific differences, while correlations/regression and principal component analysis (PCA) were employed to establish the inter-relationships between and within land and water based variables.

4. RESULTS

Five distinct land use types were identified along the Mara River based on their dominant land uses and characteristics. These were roughly forested but with human interference, agro-pastoralism, urban setting, protected/ conservancy area and natural wetland. The link between degradation and its effect on land use is central to nearly all published definitions of land degradation. In the current study, several soil physical properties were measured including soil bulk density, soil infiltration rates among others.

4.1. Soil Texture and Soil Particle Size Grades

Four different soil texture particles, namely, clay (C), sandy loam (SL), sandy clay loam (SCL) and silt clay (SC), were recorded. The current results show a decreasing trend in the proportion of clays towards the lower Mara. Proportions of sandy loam and sandy clay loam, which were negligible in the upper Mara River basin were clearly manifested at Ngerende towards the lower parts of the Mara River basin. All soil samples from Silibwet and Bomet sites showed higher properties of pure clay (C) particles, while soils from Kapkimolwa and Ngerende sampling sites recorded 3% and 24% of silt and clay, respectively. On the contrary, soils from Kirumi sampling site had higher proportions of sandy clay loam, followed by sandy loam.

4.2. Land Degradation Based on Soil Infiltration Rates

The mean infiltration over a period of 150 minutes differed, not only among sampling sites but also between different sides of the river (left and right) within the same sampling site. For instance, while mean infiltration rate on the left side of the river at Silibwet sampling site was 22.6 mm/hr that recorded on the right side was 16.2 mm/hr. The highest infiltration rate of 66.7 cm/h was recorded on the left side of the river at

Bomet sampling site and lowest 1.2 mm/hr) on the left side of the river at Ngerende sampling site (Figure 4).

4.3. Land Degradation Based on Live Biomass Levels at Different Sampling Sites

Live biomass values differed significantly between sampling sites (land use types) within the Mara River basin (one-way ANOVA, $F_{(4, 147)} = 8.57$, $p < 0.001$). Duncan Multiple Range Test (DMRT) further established that live biomass at Kirumi sampling site was significantly higher compared to all the other sites. Live biomass at Silibwet, Kapkimolwa and Bomet sampling sites did not however show any significant variation between them. Likewise, live biomass recorded at Bomet sampling site was not significantly different from that recorded at Ngerende, Silibwet and Kapkimolwa sampling sites. In addition, live biomass values recorded on the left and right sides of the river at all the sampling sites were different, with some sampling sites having higher values on one side of the river than others, as shown in Table 1.

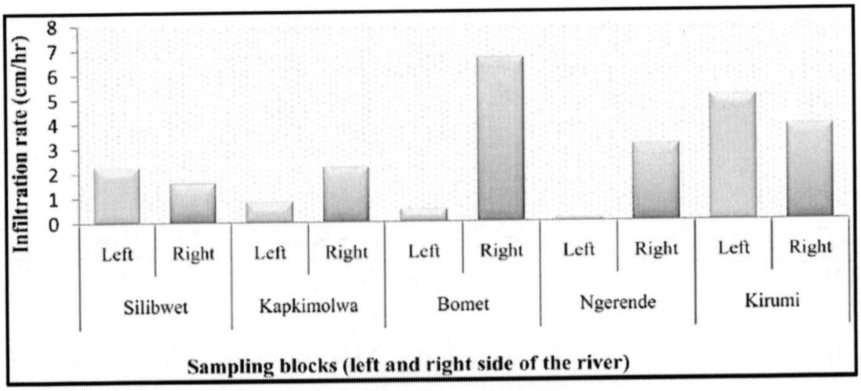

Figure 4. Infiltration rates on the right and left sides of the river per site.

4.4. Land Degradation Based on Soil Chemical Properties Due to Land Use Changes

Soil pH Levels

Soil pH was generally acidic across the five sites and varied significantly (One-way ANOVA, $F_{(4, 63)} = 19.26$, $p < 0.0001$) between sites along the Mara Basin (Table 2). Duncan Multiple Range Test (DMRT) further established that soil pH levels at Silibwet and Bomet sampling sites differed significantly from the other three sites and among themselves. Lowest soil pH levels (5.53 ± 0.43) were recorded at Silibwet sampling site and highest (6.85 ± 0.33) at Ngerende sampling site (Figure 5).

4.5. Percentage Soil Nitrogen and Phosphorus Levels

The mean percentage soil nitrogen across all sampling blocks was 4.87%, with significant differences observed in percentage soil nitrogen (ANOVA, $F_{(4, 63)} = 3.26$, $p < 0.006$) between sampling sites (Table 2). DMRT showed that percentage soil nitrogen at Kirumi sampling site differed significantly from those recorded at Silibwet and Bomet sampling sites, but did not show any significant difference with those recorded at Kapkimolwa and Ngerende sampling sites.

Table 1. Live biomass levels recorded on the left and right sides of the river.

Sampling Sites	Side of River	Live Biomass (g/m²)	Overall Mean (g/m²)
Silibwet	Left	22.63	17.14^{BC}
	Right	11.65	
Kapkimolwa	Left	20.35	21.83^{B}
	Right	23.33	
Bomet	Left	19.63	22.96^{B}
	Right	26.29	
Ngerende	Left	10.48	10.32^{C}
	Right	10.16	
Kirumi	Left	93.14	75.44^{A}
	Right	57.73	

$^{A, B, C, BC}$Means with different superscripts in the same column are significantly different at $p < 0.05$ (data analyzed by Duncan's multiple range test).

Table 2. Mean (±SD) soil parameters at different sampling blocks along Mara River.

Sampling Blocks	Soil Bulk Density g/cm³	Live Biomass	Soil pH	% Soil Phosphorus	% Soil Nitrogen
Silibwet	0.867 ± 0.12^{C}	22.96 ± 6.79^{B}	5.53 ± 0.43^{C}	1.24 ± 0.72^{B}	2.07 ± 2.20^{C}
Kapkimolwa	0.938 ± 0.10^{BC}	21.83 ± 10.41^{B}	6.55 ± 0.26^{A}	1.58 ± 0.52^{B}	5.93 ± 4.85^{ABC}
Bomet	0.881 ± 0.13^{C}	16.82 ± 7.77^{BC}	5.94 ± 0.73^{B}	1.40 ± 0.39^{B}	3.49 ± 4.67^{BC}
Ngerende	1.016 ± 0.18^{B}	10.50 ± 6.08^{C}	6.85 ± 0.33^{A}	2.27 ± 0.72^{A}	7.43 ± 6.22^{AB}
Kirumi	1.214 ± 0.19^{A}	41.53 ± 48.65^{A}	6.69 ± 0.40^{A}	2.41 ± 0.90^{A}	8.18 ± 7.94^{A}
Mean	0.956	21.204	6.24	1.65	4.87

$^{A, B, C, AB, BC, ABC}$Means with different superscripts in the same column are significantly different at $p < 0.05$.

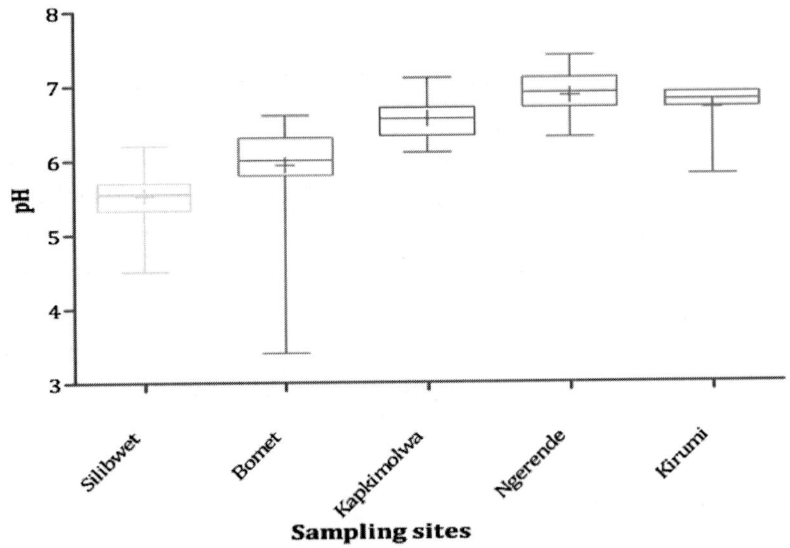

Figure 5. Box plot showing soil pH from various sampling sites along the Mara River.

Further, percentage nitrogen levels recorded at Ngerende, Bomet and Kapkimolwa sampling sites did not how- ever show any significant differences. Silibwet sampling site recorded the lowest mean soil nitrogen, (2.07% ± 2.20%) followed by Bomet sampling site-an urban area that recorded 3.49% ± 4.67% nitrogen (Figure 6).

Percentage soil phosphorus increased downstream with the highest percentage soil phosphorus being recorded at Kirumi followed and lowest at Silibwet samplign site (Figure 7).

4.6. Soil pH, N and P as Influenced by Different Land Use Types and Characteristics

Soil pH was lowest in soils obtained from pastoral land (fenced) and highest in the agropastoral (combination of crop agriculture and livestock) land which also serves as human settlement area. Percentage nitrogen was highest in agricultural and pastoral land while it was lowest on land use for controlled pastoralism (fenced). Highest phosphorus percentage was recorded in land used for agropastoral land and lowest in agricultural land and human settlement (Table 3).

4.7. Land Degradation Status Based on Soil Bulk Density

The mean soil bulk density within the Mara River basin was 0.956 g/cm^3, while there were significant variations in soil bulk density between different land use types (sites) within the Mara River basin (one-way ANOVA, $F_{(4,\ 140)} = 19.03$, $p < 0.001$). DMRT further established that mean soil bulk density recorded at Kirumi sampling site was significantly higher than all the other sites, with the highest soil bulk density (1.214 ± 0.19 g/cm^3) recorded at a grazing field at Kirumi sampling site and lowest (0.867 ± 0.12 g/cm^3) at atea farm within Silibwet sampling site. However, soil bulk density at Bomet, Kapkimolwa and Silibwet sampling sites did not vary significantly from each other, as was also the case for soil bulk density between Ngerende and Kapkimolwa sampling sites (Table 2 and Figure 8).

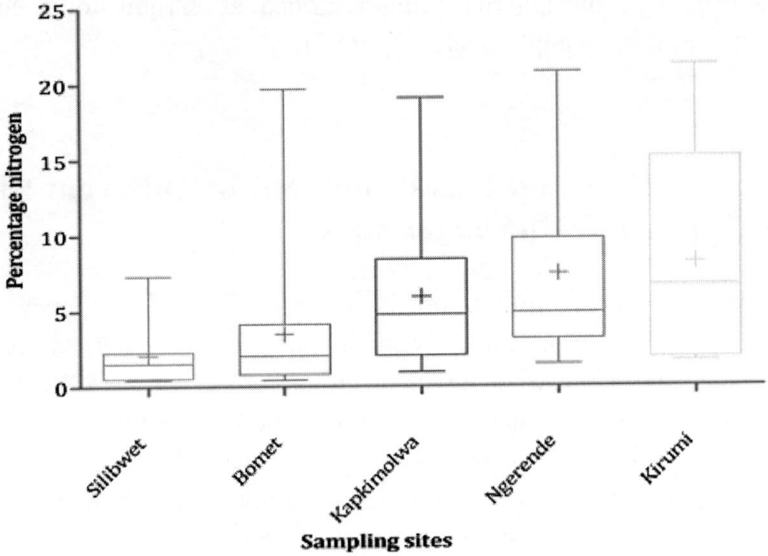

Figure 6. Box plot showing percentage soil nitrogen at various sites within blocks/clusters.

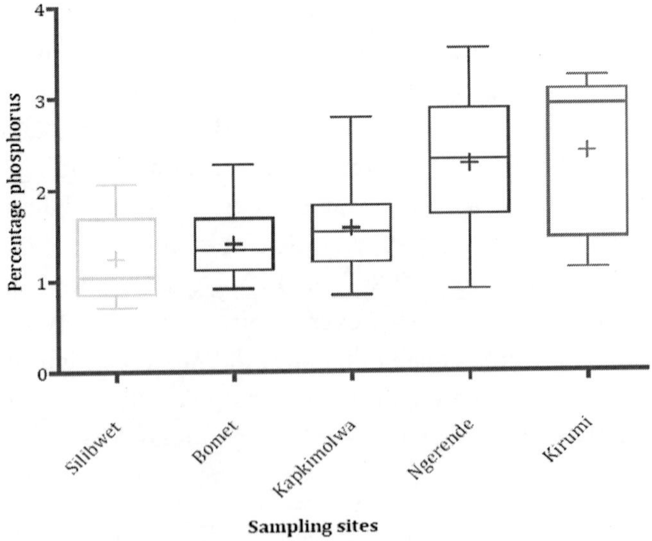

Figure 7. Box plot showing percentage soil phosphorus at different sites along the Mara River profile.

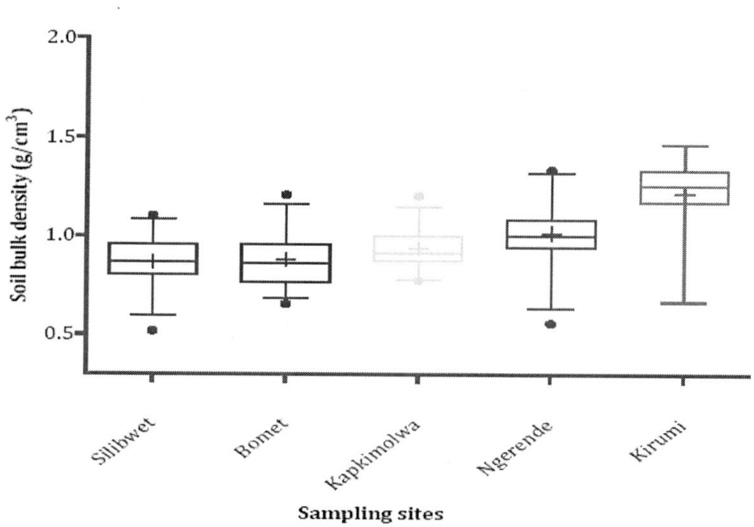

Figure 8. Box plot showing mean soil bulk density at different sites along the Mara River.

Table 3. Statistics of pH, % Nitrogen, and % Phosphorus by land use types.

Land Use Type	pH		% Nitrogen		% Phosphorus	
	Mean	s.e	Mean	s.e	Mean	s.e
Agricultural & pastoral land	5.94	0.18	0.099	1.17	0.013	0.10
Agricultural land	6.36	0.13	0.011	0.69	0.023	0.13
Agricultural land and settlement	6.43	0.13	0.018	0.64	0.010	0.66
Agropastoral land	5.33	0.09	0.005	0.09	0.024	0.27
Agropastoral & settlement land	7	0.17	0.016	0.99	0.013	0.03
Agropastoral land	6.12	0.13	0.015	0.86	0.014	0.14
Ujamaa village since 1967	5.8	0.12	0.008	1.47	0.023	0.55
Pastoral/settlement (school) land	6.4	0.19	0.004	0.74	0.011	0.11
Pastoral land (fenced)	5.42	0.31	0.014	0.03	0.012	0.08
Pastoral land (range land)	5.89	0.19	0.014	1.25	0.015	0.1

5. DISCUSSION

Characterization of land use along the Mara River profile provided a clear insight on the various land use types and rating. Mara River Basin exhibited five diverse land uses with forests (trees and shrubs) and mixed agriculture characterizing the upper Mara (Silibwet and Kapkimolwa sampling sites), agriculture, pastureland, and natural savanna characterizing the middle (Bomet and Ngerende sampling sites) and parts of the lower basin, and expansive wetlands characterizing the Lower Mara (Kirumi sampling site). Land-use change is primarily influenced by local needs, urbanization and remote economic forces. Land use change at the upper Mara River could have been triggered by the high human population growth, forcing encroachment of forest land for agricultural purposes, livestock grazing and human settlement (including urban development). Consistent with the current study findings, previous studies show that change in land use is highly dependent on the immediate needs of the inhabitants. Different land use types influence land degradation differently, with those used for agriculture and human settlements thought to cause more degradation than those used as pasture land [40] . According to Celik [41] the magnitude of land use changes vary with land use type and land management options applied. However, studies show that while hasty conversion of land use from one type to another may provide a quick remedy to the inhabitants, the long term effect on the land and by extension the aquatic ecosystem nearby may be profound.

Many researchers, including Mugisha [26] , Misana et al. [27] and Olson et al. [28] , all concur that most of the changes observed in land use in many parts of Africa, are mainly associated with extension and intensification of agricultural activities to new areas. The current findings showed that most plots within Silibwet, Kapkimolwa, Bomet and Ngerende sampling sites that had undergone changes in land use were initially forested before they were converted to farmlands, pasture lands and human settlements. One of the plots in Ngerende which was initially a forested area had also been cleared to create room for construction of a school.

Like in the current study, Maitima et al. [29] while studying the linkages between land use change and land degradation and biodiversity in East

Africa, reported that land use changes in East Africa have transformed land cover to farmlands, grazing lands, human settlements and urban centers at the expense of natural vegetation. These changes are associated with deforestation, biodiversity loss and land degradation. Studies by Lambin et al. [42] showed that land conversion to agriculture in East Africa has outpaced the proportional human population growth in recent decades. Natural vegetation cover has given way not only to cropland but also to native or planted pasture [42] . A synthesis of results of long term research by an interdisciplinary team reveals the linkages between land use change, biodiversity loss and land degradation [29] . Maitima et al. [29] reported that land use changes in East Africa have transformed land cover to farmlands, grazing lands, human settlements and urban centers at the expense of natural vegetation. They further observed that land use changes are now associated with deforestation, biodiversity loss and land degradation.

Studies by EAC [43] showed that on average 90% of the LVB population depends on crop and livestock agriculture as the main land use activity, with farm size often less than 1 hectare . This overdependence on limited land use activities contributes substantially to land degradation. Livestock and wildlife grazing has been cited as a source of soil degradation [44] . However, grazing has also been linked to increased bulk density through compaction and exposure of the soil to the sun, but reduces most soil nutrients through feeding and subsequent erosion due to the reduced ground cover [29] . One potentially degrading effect on soil condition is that of soil compaction [45] . Because soil is a complex system of biotic and abiotic components, soil compaction influences several properties of soils that may in turn affect vegetation. These include changes in root growth, availability and movement of air and water, and microbial activity [46] . Established evidence links land use change to losses of soil nutrients among them soil organic carbon, soil nitrogen and soil phosphorus, all of which reduce the productivity of the land, while adjacent aquatic systems are diminished by increased nutrient and sediment load [19] .

Studies show that in lands used for pastoral activities, soil compaction can be affected by stocking rates [47] [48] , soil texture [47] [49] , season of grazing [48] , and water content and organic matter [50] . Overgrazing of rangeland; over-cultivation of cropland; water logging; deforestation;

and pollution and industrial causes are the most frequently recognized land uses that cause of land degradation [51] . Among the key indicators of land degradation that this study comprehensively used was soil quality, characteristics and productivity. It is acknowledged that land degradation processes are not always induced by man but rather can take natural forms. In the natural state, the rate of water erosion under natural forest corresponds with the subsoil formation rate, hence there is always equilibrium. This boils to the point that accelerated land degradation commonly transpire after human intervention in the environment through different land uses.

Cultivated lands tend to have higher soil bulk densities than forested lands. However, in the current study, soil bulk density and other soil characteristics (N, P and pH) at Silibwet sampling block were the lowest compared to all other land use types within the Mara River basin. On the contrary, though they had relatively less human influence, Ngerende and Kirumi sampling blocks recorded significantly higher soil bulk density, % carbon, % phosphorus, % nitrogen as well as soil pH, compared to all other land use types (sites) probably due to wildlife and livestock trampling effects on the top soils which results in pore volumes reduction and thus high soil bulk density as well as accumulation of fertile soils from uplands to the lower reaches of the Mara River basin resulting in the high nutrient levels recorded at this section of the basin. Consistent with the current study findings, Christensen et al. [52] also demonstrated that total nitrogen and total phosphorus concentrations were highest at sampling sites with little to no agricultural activities such as Ngerende and Kirumi and lower at sampling sites that had a higher percentage of agricultural activities such as Kapkimolwa.

Soil Infiltration rates can be linked to two factors mainly the soil texture and bulk density [53] . The low infiltration rates recorded in some sampling sites such as Silibwet and Kapkimolwa could possibly be due to soil compaction from the perennial farming activities as well as due to cattle trampling effect among other factors. Studies show that soil texture controls the infiltration rate and the amount of water that can be stored in a given thickness of soil for plant use [53] . Clay soils for instance provide the highest surface area, but if clay content is great enough to restrict air and water movement, these critical variables may limit its productivity. Soils in the pure sand range on the other hand have high rates of water infiltration but are low in productivity because they do not retain water or nutrients [53] . The ideal substrate is therefore texturally balanced soil in the loam range [54] .

Land use has significant effects on soil chemical properties. Different studies have examined the effects of land use on soil physico-chemical properties [6] -[8] . The levels of soil nutrients (P and N), soil bulk density and soil infiltration rates can be used to deduce the degree at which a given site is degraded. Majule [55] and Gachimbi [56] reported that soils in areas with continuous cultivation and without appropriate management practices have low fertility levels due to overutilization. Various researchers have however reported that agricultural intensification often includes a substantial increase soil nitrogen emanating from the increased application of nitrogen (N) fertilizer, which improves yields but has deleterious consequences on adjacent aquatic systems, where nutrient loading can lead to eutrophication [57] [58] . It is however possible that nutrient uptake by crops, leaching during heavy downpours or further removal during plant harvest time could have contributed to the relatively low soil nutrient levels at Silibwet (an agricultural area) compared to other terrestrial sampling blocks downstream, while the upper ridges and urban areas like Bomet town are also susceptible to erosive nature of landscapes especially during heavy down pours which could have facilitated the active removal of top-soil nutrients especially soil nitrogen and increased erosion through runoff leaving the uplands devoid of the fertile topsoil, which are instead swept into the stream.

In this study, soil phosphorus, soil nitrogen and soil pH were assessed, to determine variations among different land use types. Studies by Maitima et al. [29] reported that grazed sites were significantly higher in soil pH and lower in bulk density, nitrogen, moisture content, percent organic matter and organic carbon than un-grazed sites ($p < 0.05$). This is a further indication that livestock and wildlife grazing activities have significant effects on soil properties with subsequent effect on water quality. The current study findings showed that soil characteristics as measured by various soil parameters including soil nutrients and soil pH varied significantly under different land use types along the Mara River.

The mean soil pH recorded in this study was 6.24, implying that the soils within the Mara River basin were slightly acidic. The soil pH of soil samples from Silibwet sampling site-which is a relatively forested site and Bomet sampling site-which was characterized by urban land use, were relatively low probably as a result of washing out of solutes from these parts as was also reported in the central highlands of Eastern Ethiopia by Mohammed et al. [59] . The high acidity at highly agricultural lands could be attributed to the decomposition and formation of carboxylic acid during the farming process. Weak acids (corresponding to vinegar) are produced in the soil when plant residues and organic matter decompose. These weak acids react and readily combine with nutrients such as calcium, magnesium, potassium, and sodium as the soil solution (water) moves down through and below the root zone (leaching). During this process, if soil pH is less than 5.2, hydrogen or aluminium replaces basic cations causing the soil in the leached zone to become more acidic [60] . Tamirat [61] also attributed soil pH of an area to the nature of the parent material, climate of the region, organic matter and topographic situation.

The relatively acidic nature of the soils could also be attributed to the high rainfall resulting in the leaching of some basic cations especially calcium from the surface horizons of the soils [62] . A pH value of less than 5.5 like that recorded at Silibwet sampling site is considered problematic for most microbial activities, and this directly influences availability of nutrients to plant [63] . Consistent with the findings of Ahmed [64] , continuous cultivation practices, excessive precipitation, steepness of the topography and application of inorganic fertilizer could also have resulted to the reduction of pH in the soil profile particularly at Silibwet sampling site where agricultural activities especially tea plantations were most dominant. Juo and Manu [65] also reported that growing vegetation tended to decrease soil pH, a phenomenon they related to cation uptake by plant, with subsequent release of H+ ions, organic matter decomposition into organic acids, increased carbon dioxide levels through root respiration and nitrification. Bobbink et al. [66] , however, reported that low soil pH can decrease plant diversity in forests especially when the soil pH is less than 4.2 making aluminium potentially toxic.

The relatively high soil pH at Ngerende sampling site could be attributed to low organic matter input in the grazing fields and probably accumulation of bases resulting from a compromised hydraulic conductivity that results from minimum leaching of the soluble bases. Simmons [67] reported that as soil pH increases above 6.5, potassium mobility slows down, and as the soil pH reaches 7.0, mobility is severely hindered making it unavailable. This is important as three of the five sites sampled within the Mara River basin recorded pH levels greater than 6.5. It was however interesting to note that high pH on land did not necessarily correlate directly to high pH levels in water. This was clear at Ngerende sampling site whereby land pH was amongst the highest but the same site recorded the lowest water pH, probably implying that pH in water was not necessarily driven or impacted by adjacent terrestrial soil pH. A number of studies show that in cropped fields, the ash deposited by the common practice of slash-and-burn releases alkaline cations (Ca, Mg and K), causing high pH, and low exchangeable Aluminium values [65] . This could probably explain the relatively high pH values recorded at Kapkimolwa, Ngerende and Kirumi sampling sites, as most studies have reported that most agricultural land is regularly limed to undo the effects of natural acidification. Studies show that many acid soils "fix" or hold phosphorus, making it unavailable for plant growth. Soil acidity can also be a barrier to root development, limiting the plants' ability to reach moisture in the sub-soil. In the humid tropics, soil acidity and associated problems often lead to land abandonment and the perpetuation of slash-and-burn agriculture [68] .

Although mean soil pH across most of the land use types in the current study were higher than the critical value of 5.5, mean soil pH at Silibwet block was much lower than the critical value. Consistent with our findings, studies show that soils from tea plantations as was the case at Silibwet tend to be strongly acidic, with some as less as 4.5 [69] . The low pH, at Silibwet sampling site corresponded with a high soil organic carbon at the same site. The long-term effect of fertilizer application could have been attributed to the low pH and high soilorganic carbon as was also observed by Wang et al. [70] in commercial tea plantations in China.

Nitrogen and phosphorus are essential nutrients for growth and development of crops, whose optimal levels in the soil is indicative of the productivity of the soil [71] . However, soils in areas with continuous cultivation and without appropriate management practices often have low soil fertility levels due to over utilization with studies further showing that major plant nutrients, e.g., potassium (K) and phosphorous (P) are the soil properties most affected by cultivation over time [55] [56] . The mean percentage soil nitrogen across all sampling blocks was 4.87%, with significant differences observed in percentage soil nitrogen ($F_{(4, 63)} = 3.26$, $p < 0.006$) between sampling sites. The continuous conversion of vegetated areas to non-vegetated surfaces as was the case at Bomet sampling site could have resulted in the reduced soil nutrients through increased soil erosion, while the significantly low nitrogen and phosphorus proportions in soils from Silibwet sampling site could have been due to crop (mainly tea) uptake as well as additional loss through food crop harvests or when vegetation is uprooted during land preparation as was also observed by Elliot [72] .

Studies show that though its consequences vary, land conversion frequently leads to nutrient losses through disruption of surface and mineral horizons (e.g. by mechanical disturbance) and reduction of organic matter inputs. Cultivation of forests, for example, diminishes soil carbon within a few years of initial conversion and substantially lowers mineralizable nitrogen [11]. Phosphorus is critical to biotic function and essential to the development and maintenance of ecosystems [73] -[75] . It is an essential element classified as a macronutrient because of the relatively important part it plays in the growth of plants. However its mismanagement can pose a threat to water quality of adjacent aquatic systems. In the current study, percentage phosphorus levels varied significantly between different sites/land use types ($p < 0.0001$) with DMRT further establishing that Ngerende and Kirumi sampling sites located at the lower Mara River Basin had significantly higher percentage phosphorus compared to the other three sites (Silibwet, Kapkimolwa and Ngerende sampling sites). Just like nitrogen, the relative low phosphorus percentages recorded at the sites located on the upper catchment particularly Silibwet, Bomet and Kapkimolwa where crop farming was dominant could have been due to removal by the crops' edible parts and plant residues. Mohammed et al. [59] linked the variability of phosphorus levels to land use, altitude, slope position and other characteristics, such as clay and calcium carbonate content. Further analysis established significant differences in percentage phosphorus level between different vegetation types.

Studies have also shown that soils devoted to crop production can lose far more phosphorus than soils that are covered by relatively undisturbed forest or natural grass land [76] . Phosphorus is an essential nutrient for plant growth hence its' active uptake by plants and subsequent removal through harvest can have an acidifying effect on the soils. Studies show that the amount of nutrients removed by cropping depends on the type of crop grown, part of the crop harvested, and the stage of growth at harvest [77] . Silibwet sampling site which is a relatively forested area characterized by small scale farms recorded the lowest phosphorus percentage compared to all the other sites, probably as a result of continuous cultivation that facilitated phosphorus uptake and removal by food crops during harvest. Livestock and wildlife herbivory can also cause shifts in plant species composition by replacing highly palatable grasses with unpalatable species [78] and therefore cause changes in soil nutrients indirectly. This is because as vegetation cover declines, soil nutrients are also depleted, while soil erosion increases, generating negative consequences on rangeland productivity [79] . Due to severe grazing, a reduction in plant biomass leads to depletion of existing nutrients among them nitrogen and phosphorus thus resulting in soil fertility reduction [79] .

Soil bulk density is highly dependent on soil texture and the densities of soil mineral (sand, silt, and clay) and organic matter particles, as well as their packing arrangement [80] . Further, bulk density is influenced by crop and land management practices that affect soil cover, organic matter, soil structure, and/or porosity. In the current study, the mean soil bulk density within the Mara River basin was 0.956 g/cm^3, while there were significant variations in soil bulk density between different land use types (sites) within the Mara River basin ($F_{(4, 140)} = 19.03$, $p < 0.001$). Most soil bulk densities fall between 1.0 g/cm^3 and 2.0 g/cm^3; while root penetration is severely impacted at bulk densities greater than 1.6 g/cm^3 [81] . As density increases, pore space decreases and the amount of air and water held in the soil also decreases [81] . In the current study, high soil bulk density observed at the lower Mara River watershed compared to other sites upstream could have been contributed by the high livestock and wild life population at those areas which together with farming activities might have had an impact on the soil structure thus increasing the bulk density and subsequently contributed to land degradation. Studies show that lower soil bulk density is desirable for plant growth, whether those plants are agricultural crops, trees, or turf grass [82] . This is because low bulk density soils have greater water infiltration rates which minimize runoff, improve water quality, and reduce storm water flow [80] .

Activities such as plowing, timber harvesting or compaction of the soil during home construction, some of which were observed at the Mara River basin can however increase soil bulk density and reduce pore space. Compacted soils may result in little to no vegetation growth in some locales [83] . The high land degradation status at the lower Mara River was also manifested in the high turbidity and nutrient concentration (i.e. total nitrogen and total phosphorus) recorded in the adjacent waters of the Mara River at this point. There was also surface runoff and soil erosion increase because these soils had lost the ability to absorb rain water. Some guidelines to minimize the damage caused by compaction include confining traffic to designated paths; reducing livestock population per unit area among others can control erosion [84] .

The significant differences in soil bulk density that existed between Nyangores and Amala and those observed at the lower Mara sub-catchments represents spatial processes occurring along the length of the river, with reduced human impact and high vegetation cover at the upland areas probably protecting soils from degradation thus leading to low bulk density. Likewise, several researchers have also reported that increased land degradation can be accelerated by the position of the land within a landscape, with lowland areas likely to suffer impacts of increased erosion, surface runoff and silt deposition, thus resulting in high bulk density [85] [86] . Silibwet sampling site which was relatively forested and with a relatively large area covered by tea plantation recorded significantly low soil bulk density, high soil conductivity and live biomass compared to all other land use types. However, bulk density recorded within Bomet and Kapkimolwa sampling sites, both located at the upper Mara River basin, were relatively higher compared to those at Silibwet.

In the current study, areas where land use types alternated between cultivated and grazing fields recorded intermediate mean soil bulk density between the agricultural crop lands and grazing fields. This is explained by the intermediate amount of litter inputs, vegetation cover and the moderate level of surface ground disturbance due to seasonality alteration of use. The ability of livestock to alter soil bulk densities is a function of stocking density and intensity of the same land use [43] . Studies show that livestock activities like simple grazing often results in soil compaction due to the weight of the animals and the mechanical forces that cattle apply when walking on land [87] . Soil compaction may in turn have negative consequences such as reduced rainfall infiltration, enhanced soil erosion [88] and degradation of the herbaceous vegetation cover [89] [90] .

6. CONCLUSION

Based on the current study findings, it was presumed that land use changes in the low land areas of the Mara River Basin took place earlier than in the uplands, as exhibited by the differences in soil particle sizes in soil bulk densities recorded at different points along the river. This phenomenon can also explain relatively low soil bulk density at Silibwet, Bomet, and Kapkimolwa sampling sites as opposed to high soil bulk density and low soil conductivity recorded in Ngerende and Kirumi sampling sites. These results point to the need to have focused policies on integrated land and water resource management strategies in the Mara River Basin.

ACKNOWLEDGEMENTS

East Africa Community-Lake Victoria Basin Commission Secretariat (EAC-LVBC) provided funds for this study. We are grateful to the Kenya Forestry Research Institute (KEFRI), Maseno, Kenya for providing the time, material, and technical support.

REFERENCES

1. Burcher, C.L., Vallet, H.M. and Benfield, E.F. (2007) The Land Cover Cascade. Relationship Coupling Land and Water. Ecology, 88, 228-242.

2. Eswaran, H., Lal, R. and Reich, P.F. (2001) Land Degradation: An Overview. In: Bridges, E.M., Hannam, I.D., Oldeman, L.R., Pening de Vries, F.W.T., Scherr, S.J. and Sompatpanit, S., Eds., Responses to Land Degradation, Proceedings of 2nd International Conference on Land Degradation and Desertification, Khon Kaen, Thailand. Oxford Press, New Dehli, India. Kaen, Thailand. Oxford Press, New Dehli, India.

3. Bai, Z.G., Dent, D.L., Olsson, L. and Schaepman, M.E. (2008) Proxy Global Assessment of Land Degradation. Soil Use Management, 24, 223-234.

4. Jones, K.B., Slonecker, E.T., Nash, M.S., Neale, A.C., Wade, T.G. and Hamann, S. (2010) Riparian Habitat Changes across the Continental United States (1972-2003) and Potential Implications for Sustaining Ecosystem Services. Landscape Ecology, 25, 1261-1275.

5. Food and Agriculture Organization (FAO) (2005) Global Forest Resources Assessment 2005. Progress towards Sustainable Forest Management. Forestry Paper 147, Rome.

6. Emadi, M., Baghernejad, M., Fathi, H. and Saffari, M. (2008) Effect of Land Use Change on Selected Soil Physical and Chemical Properties in North Highlands of Iran. Journal of Applied Sciences, 8, 496-502.

7. Agoume, V. and Birang, A.M. (2009) Impact of Land-Use Systems on Some Physical and Chemical Soil Properties of an Oxisol in the Humid Forest Zone of Southern Cameroon. Tropicultura, 27, 15-20.

8. Gol (2009) Effects of Land Use Change on Soil Properties and Organic Carbon at Dagdami River Catchment in Turkey. Journal of Environmental Biology, 30, 825-830.

9. Lake Victoria Basin Commission (LVBC) (2007) Regional Transboundary Diagnostic Analysis (RTDA) of Lake Victoria Basin. No. 4, Lake Victoria Basin Commission Publication, Kisumu.

10. Rai, S.C. and Sharma. E. (1998) Comparative Assessment of Runoff Characteristics under Different Land Use Patterns within a Himalayan Watershed. Hydrological Process, 12, 2235-2248.

11. Murty, D., Kirschbaum, M.F.U., Mcmurtrie, R.E. and Mcgilvray, H. (2002) Does Conversion of Forest to Agricultural Land Change Soil Carbon and Nitrogen? A Review of the Literature. Global Change Biology, 8, 105-123.

12. Richter, D.D., Markewitz, D., Heine, P.R., Jin, V., Raikes, J., Tian, K. and Wells, C.G. (2000) Legacies of Agriculture and Forest Regrowth in the Nitrogen of Old-Field Soils. Forest Ecology and Management, 138, 233-248.

13. Islam, K.R. and Weil, R.R. (2000) Land Use Effects on Soil Quality in a Tropical Forest Ecosystem of Bangladesh. Agriculture, Ecosystems and Environment, 79, 9-16.

14. Benoit, M. and Fizaine, G. (1999) Quality of Water in Forest Catchment Areas. Revue Forestiere Francaise, 50, 162- 172.

15. Berka, C., Schreier, H. and Hall, K. (2001) Linking Water Quality with Agricultural Intensification in a Rural Watershed. Water, Air, and Soil Pollution, 127, 389-401.

16. Wang, X. (2001) Integrating Water Quality Management and Land Use Planning in a Watershed Context. Journal of Environmental Management, 61, 25-36.

17. Liu, C.M. and Liu, X.Y. (2009) Healthy River and Its Indication, Criteria and Standards. Journal of Geographical Sciences, 19, 3-11.

18. Alin, S.R., Eilly, O.R., Ohen, C.M.C., Ettman, A.S.D., Alacios, D.L.P., Fest, M.R. and Mckee, B.A. (2002) Effects of Land-Use Change on Aquatic Biodiversity: A View from the Paleorecord at Lake Tanganyika, East Africa. Geology, 30, 1143-1146.

19. Tong, S.T.Y. and Chen, W. (2002) Modeling the Relationship between Land Use and Surface Water Quality. Journal of Environmental Management, 66, 377-393.

20. Ngoye, E. and Machiwa, J.F. (2004) The Influence of Land-Use Patterns in the Ruvu River Watershed on Water Quantity in the River System. Physics and Chemistry of the Earth, 29, 1161-1166.

21. King, R.S., Baker, M.E., Whigham, D.F., Weller, D.E., Jordan, T.E., Kazyak, P.K. and Hurd, M.K. (2005) Spatial Considerations for Linking Watershed Land Cover to Ecological Indicators in Streams. Ecological Applications, 15, 137-153.

22. Sponseller, R.A., Benfield, E.F. and Valett, H.M. (2001) Relationships between Land Use, Spatial Scale, and Stream Macroinvertebrate Communities. Freshwater Biology, 46, 1409-1424.

23. Houlahan, J.E. and Findlay, C.S. (2004) Estimating the "Critical" Distance at Which Adjacent Land-Use Degrades Wetland Water and Sediment Quality. Landscape Ecology, 19, 677-690.

24. Seitzinger, S.P., Mayorga, E., Bouwman, A.F., Kroeze, C., Beusen, A.H.W., Billen, G., Van Drecht, G., Dumont, E., Fekete, B.M., Garnier, J., Harrison, J., Wisser, D. and Wollheim, W.M. (2010) Global River Nutrient Export: A Scenario Analysis of Past and Future Trends. Global Biogeochemical Cycles, 24, 1-16.

25. Dwasi, J.A. (2002) Trans-Boundary Environmental Issues in East Africa: An Assessment of the Environmental and Socio-Economic Impacts of Kenya's Forestry Policy. Nairobi.

26. Mugisha, S. (2002) Root Causes of Land Cover/Use Change in Uganda: An Account of the Past 100 Years. LUCID Working Paper No.14, International Livestock Research Institute, Nairobi. WWW.Lucideastafrica.org

27. Misana, S.B., Majule, A.E. and Lyaruu, H.V. (2003) Linkages between Changes in Land Use, Biodiversity and Land Degradation on the Slopes of Mount Kilimanjaro, Tanzania. LUCID Working Paper No. 38, International Livestock Research Institute, Nairobi.

28. Olson, J.M., Misana, S.B., Campbell, D.J., Mbonile, M.J. and Mugisha, S. (2004) The Spatial Pattern and Root Causes of Land Use Change in East Africa. LUCID Working Paper No. 4, International Livestock Research Institute, Nairobi. WWW.Lucideastafrica.org

29. Maitima, J.M., Mugatha, S.M., Reid, R.S., Gachimbi, L.N., Majule, A., Lyaruu, H., Pomery, D., Mathai, S. and Mugisha, S. (2009) The Linkages between Land Use Change, Land Degradation and Biodiversity across East Africa. African Journal of Environmental Science and Technology, 3, 310-325.

30. ICRAF (2002) Improved Land Management in the Lake Victoria Catchment. Linking Land and Lake, Research and Extension, Catchment and Lake.

31. Mati, B.M., Mutie, S., Home, P., Mtalo, F. and Gadain, H. (2005) Land Use Changes in the Trans-Boundary Mara Basin: A Threat to Pristine Wildlife Sanctuaries in East Africa. Proceedings of the 8th International River Symposium, Brisbane, 6-9 September 2005.

32. WWF (2009) Mara River Basin Trans-Boundary Water Resources Management Programme Progress Report. Nairobi.

33. Government of Kenya (2009) The Mau Taskforce Report. Prime Minister's Office, Republic of Kenya, Nairobi.

34. USAID EA (2010) Assessing Environmental Flows for the Mara River. Nairobi.

35. Mutie, S.M., Mati, B., Home, P., Gadain, H. and Gathenya, J. (2006) Evaluating Land Use Change Effects on River Flow Using USGS Geospatial Stream Flow Model in Mara River Basin, Kenya. Center for Remote Sensing of Land Surfaces, Bonn, 28-30 September 2006.

36. Nile Equatorial Lakes Subsidiary African Program (NELSAP) (2002) Management of the Water Resources of the Mara River Basin. Project Identification No.3, 1-13.

37. Ben-Dor, E., Irons, J.R. and Epema, G.F. (1999) Soil Reflectance. In: Rencz, N., Ed., Remote Sensing for the Earth Sciences: Manual of Remote Sensing, John Wiley & Sons, New York, 111-188.

38. Hunt, G.R. and Salisbury, J.W. (1970) Visible and Near-Infrared Spectra of Minerals and Rocks. I. Silicate Minerals. Modern Geology, 1, 283-300.

39. Reynolds, W.D. and Elrick, D.E. (1990) Ponded Infiltration from a Single Ring: I. Analysis of Steady Flow. Soil Science Society of America Journal, 54, 1233-1241.

40. Greijn, H. (1994) A Missed Opportunity. Our Planet, 6, 23-24.

41. Celik, I. (2005) Land Use Effects on Organic Matter and Physical Properties of Soil in a Southern Mediterranean Highland of Turkey. Soil & Tillage, 83, 270-277.

42. Lambin, E.F., Geist, H.J. and Lepers, E. (2003) Dynamics of Land Use and Land Cover Change in Tropical Regions. Annual Review of Environment and Resources, 28, 206-241.

43. East African Community (2005) Potentials and Constraints of Promoting Lake Victoria Basin as a Regional Economic Growth Zone. EAC, Arusha.

44. Dregne, H., Kassas, M. and Rozanov, B. (1991) A New Assessment of the World Status of Desertification. Desertification Control Bulletin, 20, 6-18.

45. Soane, B.D. and Van Ouwerkerk, C., Eds. (1994) Soil Compaction in Crop Production. Developments in Agricultural Engineering Series, Volume 11. Elsevier Science, Amsterdam, 662.

46. Lal, R. (2005) Forest Soils and Carbon Sequestration. Forest Ecology and Management, 220, 242-258.

47. Van Haveren, B.P. (1983) Soil Bulk Density as Influenced by Grazing Intensity and Soil Type on a Short Grass Prairie Site. Journal of Range Management, 36, 586-588.

48. Naeth, M.A., Pluth, D.J., Chnnasyk, D.S., Bailey, A.W. and Fedkenheuer, A.W. (1990) Soil Compacting Impacts of Grazing in Mixed Prairie and Fescue Grassland Ecosystems of Alberta. Canadian Journal of Soil Science, 70, 157-167.

49. Orr, H.K. (1960) Soil Porosity and Bulk Density on Grazed and Protected Kentucky Bluegrass Range in the Black Hills. Journal of Range Management, 13, 80-86.

50. Howard, R.F., Singer, M.J. and Frantz, G.A. (1981) Effects of Soil Properties, Water Content, and Compactive Effort on the Compaction of Selected California Forest and Range Soils. Soil Science Society of America Journal, 45, 231- 236.

51. Stocking, M. and Murnaghan, N. (2000) Land Degradation—Guideline for Field Assessment. Overseas Development Group, University of East Anglia, Norwich.

52. Christensen, V.G., Lee, K.E., Sanocki, C.A., Mohring, E.H. and Kiesling, R.L. (2009) Water Quality and Biological Responses to Agricultural Land Retirement in Streams of the Minnesota River Basin, 2006-2008. US Geological Survey Scientific Investigations Report 2009-5215, USGS, Reston, VA.

53. USDA-NRCS (2001) Rangeland Soil Quality: Water Erosion. Soil Quality Information Sheet.

54. Tiedemann, A.R. and Lopez, C.F. (2004) Assessing Soil Factors in Wildland Improvement Programs. In: Monsen, S.B., Stevens, R. and Shaw, N.L. (Compilers), Restoring Western Ranges and Wildlands, USDA Forest Service, Rocky Mountain Research Station, Fort Collins, General Technical Report RMRS-GTR-136-Vol. 1, 39-56.

55. Majule, A.E. (2003) A Study on Land Use Types, Soils and Linkages between Soils and Biodiversity along the Slopes of Mt. Kilimanjaro, Tanzania.

56. Gachimbi, L.N. (2002) Technical Report of Soil Survey and Sampling Results: Embu-Mbeere Districts, Kenya. LUCID Working Paper Series No. 9, KARI-Kabete, Nairobi.

57. Beman, J.M., Arrigo, K.R. and Matson, P.A. (2005) Agricultural Runoff Fuels Large Phytoplankton Blooms in Vulnerable Areas of the Ocean. Nature, 434, 211-214.

58. Turner, R.E. and Rabalais, N.N. (1991) Changes in Mississippi River Water Quality This Century. BioScience, 41, 140-147.

59. Mohammed, A., Leroux, P.A.L., Barker, C.H. and Heluf, G. (2005) Soil of Jelo Micro-Catchment in the Central Highlands of Eastern Ethiopia. I. Morphological and Physiochemical Properties. Ethiopian Journal of Natural Resource, 7, 55-81.

60. Spies, C.D. and Harms, C.L. (2004) Soil Acidity and Liming of Indiana Soils. Purdue University, West Lafayette. www.agry.purdue.edu/ext/forages/publications/ay267.htm

61. Tamirat, T. (1992) Vertisol of Central Highlands of Ethiopia: Characterization and Evaluation of Phosphorus Statues. Master's Thesis, Alemaya University, Dire Dawa.

62. Iwara, A.I. (2011) Soil Erosion and Nutrient Loss Dynamics in Successional Fallow Communities in a Part of the Rainforest Belt, South-Southern Nigeria. Ph.D. Proposal Presented at the Staff/Postgraduate Seminar, University of Ibadan, Ibadan.

63. Solomon, D. (2008) Presentation on the Relationships Existing in Minerals Soil between pH on the One Hand and the Activity of Microorganisms and the Availability of Plant Nutrients on the Other. Bahir Dar University, Bahir Dar.

64. Ahmed, H. (2002) Assessment of Spacial Variability of Some Physicochemical Property of Soil under Different Elevation and Land Use Systems in the Western Slopes of Mount Chilalo, Arisi. Master's Thesis, Alemaya University, Dire Dawa.

65. Juo, A.S.R. and Manu, A. (1996) Chemical Dynamics in Slash-and-Burn Agriculture. Agriculture, Ecosystems Environment, 58, 49-60.

66. Bobbink, R., Hornung, M. and Roelofs, J.G.M. (1998) The Effects of Air-Borne Nitrogen Pollutants on Species Diversity in Natural and Semi-Natural European Vegetation. Journal of Ecology, 86, 717-738.

67. Simmons, J. (1998) Balancing Soil Nutrition. Rutgers Turf Management Program.

68. TropSoils (1991) Technical Report for 1988-1989. TropSoils Management Entity, Raleigh, 357 p.

69. Brady, N. (1990) The Nature and Properties of Soils. 13th Edition, Macmillan, New York.

70. Wang, R., Shi, X., Wei, Y., Yang, X. and Uoti, J. (2006) Yield and Quality Responses of Citrus (Citrus reticulate) and Tea (Podocarpus fleuryi Hickel.) to Compound Fertilizers. Journal of Zhejiang University Science, 7, 696-701.

71. Zhao, Q., Zeng, D.H., Fan, Z.P. and Lee, D.K. (2008) Effect of Land Cover Change on Soil Phosphorus Fractions in Southeastern Horqin Sandy Land, Northern China. Pedosphere, 18, 741-748.

72. Elliot, W.J. (2003) Soil Erosion in Forest Ecosystems and Carbon Dynamics. In: Kimble, J.M., Heath, L.S., Birdsey, R.A. and Lal, R., Eds., The Potential of US Forest Soils to Sequester Carbon and Mitigate the Greenhouse Effect, CRC Press, Boca Raton, 175-190.

73. Richter, D.D., Allen, H.L., Li, J.W., Markewitz, D. and Raikes, J. (2006) Bioavailability of Slowly Cycling Soil Phosphorus: Major Restructuring of Soil P Fractions over Four Decades in an Aggrading Forest. Oecologia, 150, 259-271.

74. Crews, T.E., Kitayama, K., Fownes, J.H., Riley, R.H., Herbert, D.A., Muellerdombois, D. and Vitousek, P.M. (1995) Changes in Soil Phosphorus Fractions and Ecosystem Dynamics across a Long Chronosequence in Hawaii. Ecology, 76, 1407-1424.

75. Turner, B.L. and Engelbrecht, B.M.J. (2011) Soil Organic Phosphorus in Lowland Tropical Rain Forests. Biogeochemistry, 103, 297-315.

76. Brady, N.C. and Weil, R.R. (1996) The Nature and Properties of Soil. 11th Edition, Prentice-Hall, Inc., Englewood Cliff, 740 p.

77. Whibread, A., Blair, G., Konboon, Y., Lefroy, R. and Naklang, K. (2003) Managing Crop Residues, Fertilizers and Leaf Litters to Improve Soil C, Nutrient Balances, and the Grain Yield of Rice and Wheat Cropping Systems in Thailand and Australia. Agriculture, Ecosystems & Environment, 100, 251-263.

78. Owen-Smith, N. (1999) The Animal Factor in Veld Management: Implications of Selective Patters of Grazing. In: Tainton, N.D., Ed.,

Veld Management in South Africa, University of Natal Press, Pietermaritzburg, 129-130.

79. Morgan, R.P.C. (1995) Soil Erosion and Conservation. 2nd Edition, Longman Group, Essex.

80. Sakin, E., Deliboran, A. and Tutar, E. (2011) Bulk Density of Harran Plain Soils in Relation to Other Soil Properties. African Journal of Agricultural Research, 6, 1750-1757.

81. Froese, K. (2004) Bulk Density, Soil Strength, and Soil Disturbance Impacts from a Cut-to-Length Harvest Operation in North Central Idaho. Master's Thesis, University of Idaho, Moscow, ID, 72 p.

82. Catherine, S. and Rock, P. (2007) Organic Carbon, Organic Matter and Bulk Density Relationships in Boreal Forest Soils. Canadian Journal of Soil Science, 88, 315-325.

83. Gomez, A., Powers, R.F., Singer, M.J. and Horwath, W.R. (2002) Soil Compaction Effects on Growth of Young Ponderosa Pine Following Litter Removal in California's Sierra Nevada. Soil Science Society of America Journal, 66, 1334-1343.

84. Hesselbarth, W. and Vachowski, B. (2000) Trail Construction and Maintenance Notebook. 2000 Edition, 0023 2839P, USDA Forest Service, Missoula Technology and Development Center, Missoula.

85. Lasanta, T., Garcia-Ruiz, J.M., Perez-Rontome, C. and Sancho-Marcen, C. (2000) Run-Off and Sediment Yield in a Semi-Arid Environment: The Effect of Land Management after Farm Land Abandonment. Catena, 38, 265-278.

86. Jiang, P., Anderson, S.H., Kitchen, N.R., Sadler, E.J. and Sudduth, K.A. (2007) Landscape and Conservation Management Effects on Hydraulic Properties of a Claypan-Soil Toposequence. Soil Science Society of America Journal, 71, 803-811.

87. Steffens, M., Kölbl, A., Totsche, K.U. and Kögel-Knabner, I. (2008) Grazing Effects on Soil Chemical and Physical Properties in a Semiarid Steppe of Inner Mongolia (P.R. China). Geoderma, 143, 63-72.

88. Russell, J.R., Betteridge, K., Costall, D.A. and Mackay, A.D. (2001) Cattle Treading Effects on Sediment Loss and Water Infiltration. Journal of Range Management, 54, 184-190.

89. Bouman, B.A.M., Nieuwenhyse, A. and Ibrahim, M. (1999) Pasture Degradation and Its Restoration by Legumes in Humid Tropical Costa Rica. Tropical Grasslands, 33, 142-165.

90. Alados, L., Ahmed, E.L., Aich, A., Papanastasis, V.P., Ozbek, H., Navarro, T., Freitas, H., Vrahnakis, M., Larrosi, D. and Cabezudo, B. (2004) Change in Plant Spatial Patterns and Diversity along the Successional Gradient of Mediterranean Grazing Ecosystems. Ecological Modelling, 180, 523-535.

CHAPTER 5

Soil Respiration in Drying of an Organic Soil

Zofia Fischer[1], Pavel Blažka[2*]

[1]Department of Biotechnology and Environment, John Paul II, Catholic University of Lublin, Lublin, Poland

[2]Department of Science, University of South Bohemia, České Budějovice, Czech

ABSTRACT

Drying of soil was linearly related to time, soil volume decreased and ratio of air within the soils increased. Respiration was related with decreasing humidity, storage of CO_2 in soil water results in RQ < 0.5 in the larger soil items at least for a while. Rate of drying decreased in the second part of the process. RQ increased as the CO_2 stored was aerated when its solvent-water evaporated and access of air into the soil increased; eventually RQ = 1.0 in the last days of the experiment. Respiration of the experimental soil stopped when GWC reached 0.15. ΣRQ for the whole process is about 0.7, a bit higher in experiments with less soil suggesting less anoxia.

KEYWORDS

RQ, Anoxia, Storage of CO_2, Changes of Soil Volume

1. INTRODUCTION

In the present contribution our original idea was to study process of drying more in detail. Soils in natural settings are changing; this may be approached not only in field work, but also in lab experiments, where the conditions may be set more accurately and less rely on unpredictable weather etc. and show how they are reflected in the process, particularly in soil respiration. During the work we realized however, that some concepts are generally not quite clear or understood and we propose some clarification, both by pointing to literature not frequently referred in soil science and/or through experimental design.

Respiration was/is considered as a general measure of soil activity (somewhat mystical statement, preferably of soil organisms), recirculation of nutrients from plant residues, while drying of soil is associated with wet and dry periods, seasons, hence more recently with global changes.

Soil respiration was studied many times; the literature on it is quite extensive [1] - [3] . It is based both on field measurements and on laboratory experimentation. The latter is mostly associated with sieving to get more reproducible results, and also with mixing and thus aeration of the soil material. We tried to overcome the sieving effect by slightly compressing the sieved soil by hand [4] . We have seen no other reference on the effect of compression in literature. We want to however first make some introductory comments to the subject respiration. The word respiration is historically (since 18th century) associated primarily with oxygen consumption by men and animals, more recently in biochemistry with oxidation of protons to water in the oxidative phosphorylation pathway as part of the general scheme of energy metabolism in most organisms, prokaryotes and eukaryotes; plants, animals and fungi—except only obligatory anaerobes [5] - [7] . Respiration is a heterotrophic process and also root respiration is not autotrophic e.g. [3] , [8] though it is run by organisms, which largely live autotrophically; all plants are however heterotrophs at night. Soil respiration, meaning respiration of the whole soil community, was introduced later (first half of 20thcentury—Suchtelen [9] was probably among the first). Respiration is anyway a biological process; all biological principles are to be included, not discarded, but it is also a part of soil biology, a composite science as soil physics, biochemistry, geology etc.; all parts have to be observed with the same scrutiny, accuracy and respect to achieve real progress in soil science as a whole.

The sequence/wording "anaerobic respiration" (meaning reduction of e.g. nitrate, nitrite or sulfate replacing dioxygen as proton acceptor) is used quite frequently, though it is not particularly accurate, nitrate and sulfate reduction or proton acceptance would be more to the point, but may be perhaps tolerated.

Respiratory production of carbon dioxide occurs largely in the tricarboxylic acids cycle, is associated through oxidative phosphorylation with oxygen consumption, their ratio in terms of absolute values of moles or volumes is the respiratory quotient (RQ = [production CO_2]/[consumption O_2]).

Introduction of reliable IR non-dispersive spectrophotometers (In fact already 1850 paper based on CO_2) enabling measurements of small concentrations and changes of carbon dioxide concentrations to soil science provoked some shifts in terminology; soil biological activity—measured as production of carbon dioxide—is now quite frequently referred as soil respiration e.g. [2] [8] [10] , most clearly formulated by Gaumont-Guay et al. [11] ("The efflux of carbon dioxide (CO_2) from the soil, referred to hereafter as soil respiration…"). There are however quite a few sources of carbon dioxide in soils, besides processes involving primarily inorganic substrates, e.g. acidification of carbonates by acids [12] , there is also a number of fermentation types occurring when oxygen availability is limited. Both respiration and fermentation are the only processes providing energy to organisms (as macroergic P-bonds [~P], largely ATP). Fermentation in soils contemporary with respiration is apparently quite frequent, particularly when soils become wet and both involve also consumption of some substrate(s); it is therefore surprising, if there is no information on SOM given, particularly in dry environment where very low SOM may be expected e.g. [13] . Carbon dioxide produced in fermentation is associated with much smaller gain in useful energy (about 3 to 4-times) compared to respiration. Respiration supplemented by fermentation is likely the most frequent situation in soils with inadequate oxygen availability; though even low partial tensions of O_2 are mostly adequate for microorganisms. This results in RQ ≥ 1 and decreasing production of ~P. On the other hand, many products of fermentations may be readily available substrates in respiration, if dioxygen becomes available again. This is also reflected in the interpretation of the RQ. Very often limits for RQ are given e.g. within 0.7 till 1.3. In fact in anaerobes (obligatory or facultative, microbial or eukaryotic) RQ may be much higher (e.g. >10), in strict anoxia it cannot be defined at all, as dividing by 0 (O_2 consumption) is a prohibited math operation. Two most important types of fermentation for soil may be lactic and ethanolic (alcoholic). While respiration of lactate and many other products of fermentation (including e.g. acetate) results in RQ = 1, like most carbohydrates; for catabolism of ethanol RQ = 0.57; in reduction of methane RQ = 0.50. Efflux of CO_2 is in fact a mix of respiration, fermentation and abiotic processes leading to absorption or release of CO_2. Field measurements of CO_2 release to the atmosphere are useful particularly for understanding of the role of soils in the global carbon cycle. Experimental conditions (e.g., participation of autotrophs, soil humidity, differences of temperature between the "dome" and the neighboring free area...), are however to be thoroughly specified and the CO_2 efflux is not considered as soil respiration.

In soil literature retention of CO_2 in soil water due its high solubility (distribution coefficient air/water = approx 1 compared to dioxygen 0.03) is mentioned only exceptionally. In soils with pH > 4 dissolved bicarbonate becomes the dominating form of the carbonate system [14] . What we measure by any instrument is the apparent production of CO_2 and RQ modified by diffusion and solubility interference. RQ is a ratio, relationship between the differences of the two components and their ratio are thus non-linear, rather exponential. Jenkins and Adams [15] are thus not correct in saying "RQ is free of empirical assumptions...", Araujo et al. [16] in accepting without comment measured RQ in the range RQ = 0.2 till 1.5 depending on the soil degradation level. On the low side remediation of hydrocarbon degradation of soils results in low RQ [17] , again below the theoretical limit 0.65 for hexadecane; methane is the most reduced hydrocarbon, its oxidation (assuming only one process) results in RQ = 0.5; this should be considered as the low limit of acceptable RQ data. On the other side of possible range synthesis of lipids from carbohydrates are sometimes considered with RQ ≈ 1.3 as possible maximum. It is probably correct for respiration; we must however expect hypoxic or anaerobic conditions particularly in wet soils. Soil organisms are forced to get additional energy to what they get from decreased or even stopped respiration through some fermentation; with prevailing fermentation RQ may be well over 10. In deep anoxia RQ cannot be calculated at all, as CO_2 is produced, but there is no oxygen consumption, division by 0 is a prohibited math operation. Let us just remind summed equation of ethanolic fermentation:

$$C_6H_{12}O_6 = 2\,CH_3 \cdot CH_2OH + 2\,CO_2$$

Being aware of interferences in measurements is particularly important in studies of processes, like various kinds of acclimation to changing temperature and humidity or in the budget of a process.

We want to analyze particular parts of the drying process, not only desiccation, but also changes of soil volume, access of air/oxygen to soil aggregates, storage capacity for CO_2 within the soil etc. and their impact on measured respiration data. This applies both to laboratory and field setups and involves size of soil items in both. Our soil is to some extent exceptional in high SOM content (around 50% of dry weight), many relationships studied in less organic soils need not to be applicable, also some of our conclusions are necessarily not of general use.

Drying of soils is a natural part of regular seasonal events, in observations and/or experiments many papers tend to limit themselves to just momentarily flashes or steady states; we tried however to estimate the budget of the drying process to go through different situations, states of the soil, get rid of many obstacles in getting real metabolic data. Incidentally, we came across low rates, their statistical treatment and time scales.

2. MATERIAL AND METHODS

2.1. The Site and Soil Sampling

The location is on the northern outskirts of the city of Warsaw (Poland). Geographic coordinates are: 52°20'212 N, 020°51'260 E, elevation 81 m. Refer to Blazka and Fischer [18] . The aim of the study is understands of mechanisms of drying through laboratory experimentation and not description of a particular locality. We used the soil for experiments as a laboratory model. The soil is highly organic; nearly 50% LOI, most of the mineral fraction is sand. It comes from a wet site in a dense forest, from the very active A1 zone, depth about 10 cm below surface, it was 5 mm sieved. We got our samples respecting suggestions by Carter and Gregorich [19] .

2.2. Laboratory Analyses

2.2.1. Humidity

(GWC—gravimetric water content, WHC—water holding capacity, dm—dry mass).

Our procedure was based essentially on Grace et al., Priha and Smolander [20] [21] with minor modifications [18] . GWC and other contents are given as ratios, but not %. Dry mass was determined by drying at 110°C to constant weight.

2.2.2. Soil Organic Matter (SOM and Density)

SOM was estimated as loss of weight in ignition to constant weight (LOI) at 500°C on selected samples. Mean densities of SOM ($\rho = 1.3$) and of the mineral fraction of dry weight ($\rho = 2.65$), are from Ilsted et al. [4] and many others.

Organic matter = LOI was between 0.45 and 0.50.

$\rho(\text{soil}) = (\text{LOI} * 1.30 + (1 - \text{LOI}) * 2.65 = 1.96.$

2.2.3. Respiration

(OCD—oxygen consumption, PCO_2—production of CO_2, RQ—respiratory quotient).

We used the volumetric method with the Scholander type respirometer to get both the oxygen consumption and carbon dioxide production. The capacity of its cylindrical flasks is 50 ml (inside radius r = 2 cm, V = 4 cm^2) and we refer data as $\mu l \cdot g^{-1}$(dry mass)$\cdot h^{-1}$. Each experimental series consisted of two parts. In the first, without hydroxide in the flasks, changes in the volumes were recorded mostly for 5 hours, this gave the sum of CO_2 changes and O_2 consumption, in the second part after placing filter paper soaked with NaOH solution into the respirometric chamber, we recorded the oxygen consumption (OCD) for another 5 hours interval, readings each 30 minutes. This involves assumption of slow rates of metabolic changes. The bath temperature was 17.5°C. Sensitivity of the volumetric respirometer is effectively not much worse (<0.1 μl $gas \cdot g^{-1} \cdot hour^{-1}$) than of the IR meters, though rates of providing data are not comparable.

While data of the second part were the OCD, PCO_2 was calculated as first reading (without hydroxide) + second reading (with hydroxide), respecting the signs in readings and calculations. We then plotted the results against time of readings, the slope coefficient of the regression (β_1) was the rate (O_2 consumption-OCD or PCO_2); the respective regression coefficients were all $r > 0.95$, they were later corrected for pressure and temperature, multiplied by the calibration factor and related per gram dry soil and hour. Weights of respiration vials with soil were taken regularly; tare was subtracted to get rates of drying.

Later in Results and discussion we will consider to what extent the data measured might be biased by diffusion or other transport problems within the soil items, the measured data may then be classified as apparent data (AOCD, $APCO_2$ and ARQ).

2.2.4. Soil (Vs), Air (Va), Pore Volume (Vpo)

Wet soil was gently pressed before start, there were no fissures between soil and wall of respirometric flask, they appeared however later during drying. We measured them by a series of blades (used in fine mechanics to measure narrow crevices, range 0.05 till 2 mm) on four sides of the vial, their mean was subtracted from r = 2 cm to get the radius of the base. The height (he) of the cylinders was estimated using a plate put on the top of the open vial and fixed to center by 4 screws projecting down on the outside of the flask, there were 5 evenly distributed holes on the plate, through which the distance from the top of the plate to soil top was measured by a rod, read by caliper and subtracted from the distance bottom of flask-top of plate. Mean of the 5 readings was height of the soil (he). Final calculation as cylinder volume:

$$\text{Soil volume } (Vs): Vs = \pi * r^2 * he \tag{1}$$

$$\text{Pore volume}(\text{porosity})\, Vpo = Vs - ((Vs - Ws)/\rho) = Vs - ((Vs - H_2O)/\rho) \tag{2}$$

Ws is mass of wet soil.

$$\text{Volume of air } Va = Vpo - H_2O \tag{3}$$

2.3. Experimental

Changes of GWC in experiment are included as background for other parameters of the soil studied. We refer four principal experiments, which were preceded by several preliminary checks or pilot experiments, all done on one and the same kind of soil collected from one spot. All were aimed towards analysis of changes in soil respiration (OCD, PCO_2 and RQ) in drying of the soil. The main difference between the experiments was in the amount of soil in flasks and therefore were coded as LS (less soil) and MS (more soil) respectively. Outline of the experiments is in Table 1. Drying proceeded in MS1 and MS2 in two rates, up to about half of the experiments the rate (a) was somewhat faster, then later (rate b). Drying was terminated when oxygen uptake and CO_2production dropped below the sensitivity of the instrument, except LS1 terminated earlier. There were 12 (LS1, MS1) or later 6 (LS2, MS2) parallel respirometers in each series. The basic data differ by an order of magnitude and change during the experiment; therefore we compared the coefficients of variance-cv's. It turned out that the cv's of the means based on six replicates are smaller than those of 12 replicates (Table 2). As standard deviations (second roots of variance) are involved in calculations of the t-test, we consider the data as comparable (Table 2). Weights of vials for GWC estimates were taken not necessarily on the day of respiration measurement. Therefore GWC data for comparison with respiration were interpolated from the respective regression of GWC to time (all $r > 0.99$).

After each 10 days WHC was estimated in 10 of the vials in the first series (MS1, LS1). In experiment MS2 and LS2 we estimated WHC at the beginning and at termination of the experiment.

Additionally we made a 14-days drying experiment (MP—mud pies) starting at GWC = 1.0 and finished at GWC = 0.13 at the respiration stop. There were two modes of soil drying, each in six respirometers: a) soil (350 g dry weight) dried freely distributed on a plate, aliquots were taken for each respirometric measurements (per 6 * 5 g dry weight), returned to the plate and mixed; b) In the second mode wet soil was pressed by hand into the respirometric vials (per 5 g to each vial), it was left to dry freely between measurements each second day.

We also checked respiration in strict anoxia (SA) comparing respiration of a fresh collected soil with respiration after flushing the respirometers by pure dinitrogen (Air products.com, purity 5.2).

Table 1. Outline of the experiments.

	LS1	LS2	MS1	MS2
dry mass [g] mean; se	3.8; 0.0294	3.26; 0.025	15.22; 0.0751	14.61; 0.147
days of experiment	24	13	52	42
starts of experiment	Aug 25. 14	Feb 23. 15	Nov 11. 14	Feb 23. 15
1st linear section [d]	24	13	32	29
°C air	17.2 - 19.0	18.8 - 19.7	17.0 - 18.8	18.4 - 19.7
β_1 GWC rate a [m]	−1.845	−4.43	−1.632	−1.257
β_1 GWC rate b [m]	x	x	−0.721	−0.412
mud-pies formation [d]	8	6	16	20
humidity air %	50 - 65	40	45 - 62	40 - 45
GWC start	2.4	1.54	2.34	1.58
GWC end	0.644	0.12	0.15	0.16
GWC estimates N	7	5	15	15
GWC replications	12	6	12	6

GWC rates are β_1-coefficients in linear regressions per month. a stands for the first, b for the second linear section. In the LS was one section. LS1 was finished prematurely.

Table 2. Cv's of means of cv's in MS1 and MS2 experiments for OCD, PCO_2 and GWC to compare variation in data with 12 or 6 replicates.

	OCD	PCO_2	GWC	replicates	determinations
MS1	0.2919	0.3581	0.1275	12	14
MS2	0.1373	0.1401	0.0511	6	13

2.4. Statistics

Was rather basic, it involved averages = means (m), standard deviations (s), coefficients of variation (cv) and standard errors of the mean (se) calculated for all series of data obtained, referred here or not. Two sided t-test was applied for checks of differences of starting and closing data of series and for checks of differences in neighboring data within series. In final evaluation principal (Table 2) rates were related to month: (30days per month).

This came out from the discrepancy in evaluation of rates in changes in GWC, Vs, Vpo related to day which was our working time unit. We were unable to reject the 0-hypothesis (slope-rate of change of GWC is not different from zero—(from horizontal plane line), whereas paired T-test showed significant differences not only between start and end of experiment ($p < 0.001$), but also between most of neighboring data in the series (Table 3). In general the choice of reference units data is frequently intuitional, comparison of two or more statistical treatments might be a more general tool.

3. RESULTS AND DISCUSSION

Results are largely based on the four drying experiments specified in Methods, Experimental, Table 1 and in Figure 1. Additionally we performed two checks to get respiration data of the hard mud pies and the small sandy like aggregates and demonstrate CO_2 production in strict anoxia. Also we assured us that the hard dry mud pies are not just gypsum or carbonate.

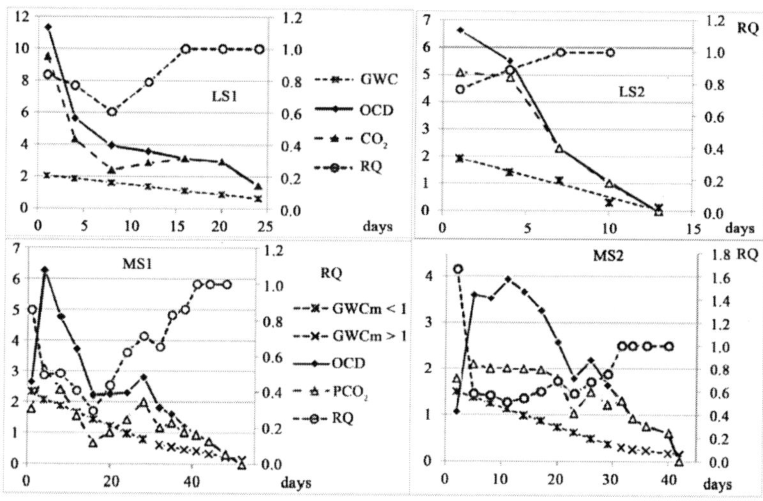

Figure 1. Changes of the gravimetric water content-GWC (as ratios), OCD (oxygen consumption), PCO_2 (production of carbon dioxide)—both as $\mu l \cdot g\ [dm]^{-1} \cdot h^{-1}$, and RQ (alternative scales) in the four experiments. Rates of changes in GWC are in Table 1 and Table 3, rates of OCD and PCO_2changes as related to GWC are in Table 5, overall budget of the experiments in Table 6. The LS1 experiment was finished before respiration dropped to zero. All standard errors are within the point icons. Specification of symbols for LS2 and MS2 experiments are in parts for LS1 and MS1 respectively.

Table 3. Rates (slopes) GWC, Vs, OCD per week and month for MS2 and LS2 compared to 2-sided t-test for start-end and neighboring days.

MS2	β_1 (slope)/day	N = 14 all p	slopes/month	p	T-test of pairs start-end	T-test of neighboring days			
						$p > 0.05$	$p < 0.05$	$p < 0.01$	$p < 0.001$
GWC	−0.0361	>0.05	−1.083	<0.005	$p < 0.001$	0		0	14
Vs	−0.4255	>0.05	−12.76	<0.001	$p < 0.001$	6	5	1	2
OCD	−0.0733	>0.05	−2.319	<0.001	$p < 0.001$	3		1	10
LS2		N = 5 for GWC and Vs; N = 4 for OCD							
GWC	−0.1138	>0.05	−3.41	<0.005	$p < 0.001$	0	0	0	5
Vs	−0.3963	>0.05	−11.89	<0.001	$p < 0.001$	0	0	2	3
OCD	−0.5904	>0.05	−17.71	<0.001	$p < 0.001$	0	1	0	3

Significance of slopes related to day and month, compared to 2-sided T-test for data pairs start-end of experiment and neighboring pairs during the experiment.

3.1. Humidity

Relationships of humidity (GWC) are therefore the first to be examined more in detail. Its changes related to time in terms of days[d] proved to be controversial; this resulted in use of month [m] as time reference unit in some selected relationships (Table 3).

GWC decreased during the experiments linearly with time for approximately 2/3 of the course, towards the end in the MS experiments, there was a significant drop in rate (Table 1). The course was a single linear function with respect to time in the shorter LS experiments. A slowing of the function is indicated in LS2, there are, however, too few data to get a significant picture (Figure 1).

3.2. Soil Volume

So far cracks in dry soils were often referred from the field in the literature, but there was/is much less interest in experiments on changes of soil volume in lab. There are more recent works on the impact of repeated drying and rewetting cycles on characteristic of soils resulting so far in non-complete understanding of these processes. We registered shrinking of soil during LS1 and MS1 experiments and estimated it with assumptions of linearity; nevertheless these estimates for MS1 and LS1 are in good agreement with the data from the second series of experiments. All four experiments have shown significant shrinking and changes of pores (porosity). The LS2 experiment is less successful due to high rate of drying and GWC decrease, therefore it was short and only few data were obtained (Table 2 and Table 3).

Soil humidity (GWC) appears to be the governing parameter in shrinking of soil and respiration of soil biota. Shrinking [soil volume Vs, cm^3] and volume of calculated pores, both Vpo and Va, are linearly related with GWC, only Va in LS2 is not significant (Table 4). Regressions of Vs to time [d] were significant.

3.3. Changes of Soil Consistency

Formation of "mud-pies", hard stone—like concretions of dried soil, occurred in all four experiments. In MS1 and MS2 it begun approximately in 1/4 of the drying time, resulted from shrinking of soil, fissures appeared between soil and the walls of respirometric flasks, thus more of the soil surface was opened to the gas phase in Va did not change significantly in drying, ratio to Vs changed just due to decrease of Vs. the flasks. The rate of drying was also related to the amount of soil in the flasks. In the MS1 and MS2 series there is a higher soil layer, longer diffusion path and supposed humidity zonation. In the LS series drying and course of the experiments were faster (Table 1); the "mud pies" eventually formed just a narrow plate of dried soil.

All these processes result in increasing access of air into the soil preparation, in a much better ventilation of the dried soils than is in the wet soils and contributed to formation of the "mud pies" during the drying experiments.

It's a common knowledge, that wet soils are soft, pastry; easily adopt form of container under only very slight pressure. When dried they become hard, stone-like, if slightly pressed together before drying; they resist pressures approximately like wall nuts, larger "mud pies" are more resistant than small ones. Sieved soil loosely distributed on a plate acquires in drying a sandy character; its total surface is much greater than that of the "mud pie".

Table 4. Soil volume (Vs), porosity (Vpo) and air filled porosity (Va), their % share in Vs, rates of change (β_1, r), probability of no change (p).

	LS1	LS2	MS1	MS2
Vs; start	9.79 (d8)	8.042	37.7 (d16)	32.735
N; se	9;	9; 0.33	13;	24; 0.509
Vs end	8.17 (d20)	3.035	25.4	15.987
se	x	0.222	x	0.245
Vs-GWC; β_1	1.57	3.74	7.45	11.73
−r	0.944	0.977	0.941	0.990
$p<$	0.025	0.001	0.001	0.001
%Vpo in Vs start	84.2	83.2	87.5	86.38
%Vpo in Vs end	62.3	56.5	50.1	58.32
Vpo-GWC; β_1	2.1775	3.15	10.696	13.252
−r	0.955	0.9947	0.995	0.8767
$p<$	0.001	0.001	0.001	0.001
%Va in Vs start	15.18	16.13	10.3	15.13
%Va in Vs end	36.20	41.79	45.01	68.75
Va-GWC; β_1	−1.13	−0.139	−3.572	−1.468
−r	0.983	0.347	0.964	0.877
p	<0.001	>0.05	<0.001	<0.001

Va did not change significantly in drying, ratio to Vs changed just due to decrease of Vs.

To check the character of the hardiness of the mud pies we chopped off 5.0 g material from the top of the 25 g "mud pie" and took an aliquot of the same weight from the sandy like dried soil. Both were suspended in 50 ml of pure water and extracted for 2 hours and 30 minutes. Both soils looked hydrophobic at the beginning, after approx. 45 minutes they turned more polar. Some air bubbles were attached to dark, presumably light organic soil particles. The extracts were separated from the rests of soils by filtration through Macherey-Nagel filter paper MN 640 m. Both filtrates were yellowish brown, the "mud pie" extract was distinctly darker, but not brown, both very slightly colloidal. Most of the residues remained in the Erlenmeyer's with the heavy sand fraction, much less on the filters. We measured pH, conductivity and redox potential in both extracts: pH = 4.48 and 4.53, conductivity = 54 and 116 $\mu S \cdot cm^{-1}$ in "sand" and "mud pie" respectively, redox was E = 320 mV in both. Addition of 1 ml 0.2 molar solution of $BaCl_2$ to each of the extracts did not result in formation of any precipitate. We conclude, that the top of the "mud pie" probably contained more organic matter than "the sandy" fraction (more intense humic coloration of the extract), there was certainly no carbonate nor gypsum formed in the drying soils, could not been responsible for formation of the "mud pies". The two major components of our soil are the postglacial sand, which however remained unchanged in all the procedures as a separate and non-reactive fraction. The second major component is SOM.

We want suggest replacement of H-bonds (water bridges) in wet soil by some other weak interactions (non- covalent bonds) during drying, possibly by hydrophobic weak bonds. A slight pressure and greater mass of wet soil brings soil aggregates closer together and enables formation of other particle interactions when O-(water-) bridges are abolished in drying.

For considerations of bonding within soil aggregates a change in scale of the aggregates might be useful. Remind that there are 6E23 single molecules per each mole (Avogadro number), thus in 1 μg of SOM 6E14 molecules of organic compounds of mean molecular mass 1000 may be accommodated. Soil science does not seem to be generally ready for thinking in such scales (unfortunate Table 2.1 in Voroney [22] , containing organic molecules [10^{-9} m] larger than viruses, $\Psi > -1500$ kPa not only in Chowdhury [13] etc.).

3.4. Respiration

3.4.1. Relationships to GWC

Relationships of gas exchange (OCD, PCO_2) with time and GWC are more complex, the respective regression coefficients are lower compared to regressions of GWC and time; the overall tendency of respiration (OCD and PCO_2) is a decrease with progressing drying as water is a primary essential-even substrate-of life (Table 5, Figure 1). Data on respiration, both OCD and PCO_2, follow common trends: high values immediately at the beginning of experiments in both LS, or after three or five days in the MS series. There is a more or less stable respiration in the central part of the course in MS1 and MS2 (Table 2 and Table 4). Basic statistics of respiration data show small variance, se's in all presented and vast majority of data not shown are within the point icons.

The data in figures are of similar magnitude with those published by selected authors for forest soils considering problems in recalculation of area related data to mass related. Butler et al. [8] consider "rainfall as main driver of soil CO_2 efflux and its components".

Table 5. Regression coefficients (β_1, r) for respiration to GWC.

r/GWC	LS1	LS2	MS1 all	MS1 after d2	MS2 all	MS2 after d3
OCD β_1	−5.104	−3.612	−1.970	−2.560	−2.000	−2.950
r	0.833	0.956	0.837	0.936	0.712	0.949
PCO_2 β_1	−3.704	−0.877	−0.816	−0.978	−1.149	−1.351
r	0.735	0.953	0.732	0.765	0.842	0.881

For MS experiments coefficients are for all data and also for dates after d2 and d3 due to the delayed Birch effect.

3.4.2. CO_2 Storage

Distribution coefficient between gas and water for CO_2 is very close to one at the temperature of the experiments, storage capacity for CO_2 in soil water is therefore equal to GWC * dry mass of soil. It is the volume of water in respirometer (microcosm), about 20 ml in the first days of the MS experiments or 5 ml in the both LS; apparent production in two or three days.

Another estimate of CO_2 stored may result from the difference between AOCD and $APCO_2$ during the flush of respiration when soil is still wet, in MS1 days 3 to12. Assuming real RQ = 1.0 it is 2.81 ml CO_2 stored in 28.8 ml H_2O, the storage capacity is thus used in less than 10%.

Similar data for the MS2 experiment is for days 5 to 12 and show 2.79 ml CO_2 stored in 12.7 ml water, storage capacity used in approximately 22%. Assuming real RQ = 0.7 result in 1.97 ml CO_2stored in MS1 and 1.95 ml CO_2 in MS2 respectively. Details of calculations are not shown.

The storage capacity for CO_2 was falling in experiments with progressing drying. Angert et al. [23] compared measured data of RQ (ARQ) with soil air analysis showing relatively low pCO_2. Their assumption on equilibrium between headspace (soil air) and wet acidic soil might not be fulfilled; there might be physical separation between the sections of the soil gas and the watered section, where CO_2 is primarily produced. Also a careful acidification and mixing of a tightly closed dilute solution of bicarbonate without gas phase need not result in production of gas bubbles. This may apply to a wet soil, where CO_2 is produced by microorganisms. ARQ data of Angert et al. [23] agree with our RQ in the respiration budget, there is also agreement on denial of carbonate chemistry as cause in the background of RQ < 1, we propose oxidation of methane and particularly ethanol produced in fermentation as hypothetic substrates in respiration with RQ < 1. We also seem to agree in accent on processes in studies. Sierra and Renault [14] conclude, that production is negligible compared to dissolved CO_2, this does not apply to acidic soils (pH < 4), where bicarbonate is practically missing. Therefore we did not consider bicarbonate storage, as estimates in our soil were always pH $<$ 4.0. In acid soils however even a small volume stored may result in serious shifts in RQ.

3.4.3. Anoxia in Soil

To show gas exchange in strictly anoxic soil we placed 21 g fresh collected sieved wet soil (7.3 g dm) into each of four respirometers. Two of the respirometers were flushed with pure dinitrogen 50 $ml \cdot min^{-1}$, for 10 minutes; we registered O_2 consumption and CO_2 production [with and without NaOH trap inside]. The other pair of respirometers was a control to see regular soil respiration. There were 3.83 and 3.89 $\mu l \cdot g^{-1}(dm) \cdot h^{-1}$ respectively gas produced in the N_2 flushed respirometers without the hydroxide trap. There was however no gas production, nor consumption when hydroxide trap was placed in both respirometers; this also shows that the gas produced without the trap is absorbed by hydroxide, it is most probably CO_2 and not CH_4. RQ cannot be determined as shown in Introduction. The regular respiratory control in the second pair of respirometers provided 8.92 and 8.56 μl $O_2 \cdot g^{-1}(dm) \cdot h^{-1}$ respectively and 4.51 μl $CO_2 \cdot g^{-1}(dm) \cdot h^{-1}$ in both respirometers, this is consistent with data in Figure 1.

3.4.4. Respiration and Size of “Mud Pies”

Another pilot experiment aimed to differences of respiration in highly dried soils as “mud pies” or small sandy like aggregates. In the “mud pies” both OCD and PCO_2 are some 20% lower than in the sandy sieved small aggregates. Fast drops of simultaneous decrease of OCD and PCO_2 by approximately 0.6 - 0.8 $\mu l \cdot g^{-1} \cdot h^{-1}$ occurred at 0.63 GWC in “mud pies”, at 0.43 in “sand”. Drying in “mud pies” and associated decrease of respiration is slower in “mud pies”. There is no contradiction to Figure 1.

3.4.5. “Birch Effect”

High respiration data occurring after changed conditions (sieving, watering…) are referred as the Birch effect [24] . In our experiments it occurred in LS1 on d1, in MS1 on d 5 - 14 and also at lower GWC = 1.41 starting value in LS2 and MS2 thus at a somewhat lower moisture throughout the drying experiment (Figure 1). The delayed respiration flush in both MS experiments resulted in distorsions of the relationships of respiration to GWC, this was overcome by skipping the d 1 or d 2 data from the relationships in MS2 and MS1 (Table 6). The delay of flush in both MS experiments was probably caused by pressing the greater amount of soil at start resulting in formation of solid “mud pies”, while the small amounts of soil in both LS resulted in a thin layer of soil with minor limitation in gas traffic in the first days. The slightly smaller humidity in MS2 compared to MS1 was of minor importance in this range. We however feel that the impression of extraordinarily of high starting RQ’s in MS2 is partly misleading, it comes rather from general characteristics of ratios; OCD and PCO_2 data are of similar magnitude (1.78 and 1.07 respectively), not strikingly different indeed.

High flushes of respiration following change of conditions are debated for last 50 years [24] . So far they were attributed to increased availability of respiration substrates from organisms which died in preceding dries, even if the drop in water content was small [25] or to repeated drying and rewetting (e.g. [26]) or simply to watering of soil (e.g. [13]). Even the detailed study by Unger [27] which includes interaction of soil and the vegetation using highly advanced methodology does not seem to be the last word in the debate. Flush took place in all four experiments in fresh and sieved samples, but not watered, of course sampling was at the very beginning (Figure 1, Table 5, first section of the course). Respiration decreased dramatically in the next readings. Therefore we want to propose a hypothesis not relying primarily on the experimental conditions, but rather on the organisms themselves: Before the collection of soil organisms were adapted to low respiration rates in their natural settings, supplemented by fermentation. Sieving (or in other papers published-water added) allows more air to enter the soil; previously accumulated CO_2 is released to air or absorbed in water. The overall tendency during the experiments is decrease of AOCD and $APCO_2$ with decreasing GWC, the data of the first part of experiments are however influenced by CO_2 produced and stored in soil water and Va available, at this stage it was not included in the changes of gas contents measured. Some loss of humidity (GWC) in sieving is highly probable. Soil biota may require time to adapt to available dioxygen, presumably by activation/reconstruction of respiratory mechanisms (e.g. enzyme complexes of oxidative phosphorylation pathway) which appear to be of small use in more enzyme copies in wet soils in natural settings with limited oxygen availability.

3.4.6. Summing up OCD and PCO_2 in the Process

Summed OCD and PCO_2 for the whole drying experiment showed in all four experiments RQ around 0.7 (Table 6). ΣRQ is somewhat higher in the faster drying experiments (LS), suggesting less hypoxia. This may support the hypothesis presented in introduction on a mixed status of organisms in wet soils, respiration contemporarily with fermentation and use of the fermentation products (e.g. ethanol) as substrates in respiration when O_2 becomes available.

Studies of some process which take into account all data, characterizing the variable soil condition during a process may provide more information, than short measurements or flashes of soil metabolism in ± constant conditions. Type of instrumentation may be of secondary importance only?modern methods of O_2 estimation are used only exceptionally [27] .

Nevertheless if there is O_2 consumption, anoxia is not in the total mass of soil, possibly in a limited area in center of the soil bloc.

3.4.7. Respiration Ends at Limiting GWC

There is a small, but significant peak of respiration in at the beginning of the decreasing phase at d 26 in MS1; in MS2 at d 26. A similar peak was registered by Thomson et al. [10] on the 5-th day of their 7-days experiment, which did not continue to a respiration stop. We propose to interpret as increased respiration due to increasing access of air to the drying soil, in our experiments shortly overrun but proceeding desiccation of the soil and confirmed by fast drops afterwards.

Table 6. Sums of OCD, COP, and ΣRQ in four experiments [ml].

	LS1	LS2	MS1	MS2
ΣOCD	7.67	15.14	42.53	33.13
ΣCOP	6.31	13.14	23.97	22.41
ΣRQ	0.82	0.87	0.56	0.68

ΣRQ is the ratio of the two sums. Duration of experiments in Table 3.

All four series (including the prematurely finished LS1) end up with RQ = 1 when GWC is approximately = 1 in LS experiments, but GWC ≈ 0.6 in MS (Figure 1).

CO_2 retained in the first section of experiments in water is completely aerated from the soil towards the end when soil is very dry; when respiration seizes at GWC ca 0.15, there is no measurable PCO_2, as is no OCD.

All drying experiments were terminated when GWC dropped to 0.15 and respiration stopped. The experiment LS1 was in fact finished prematurely, nevertheless its final RQ = 1. Also in a preliminary experiment not reported here (summer 2014) respiration (both OCD and

PCO_2) seized at GWC = 0.15 and RQ = 1. The respiration limit at GWC = 0.15 need not be a universal value for many soils, Chowdhury et al. [13] reported respiration in Australian soils at recalculated GWC = 0.025.

4. CONCLUSIONS

There is a distinct peak of respiration rate at the beginning of the experiment (Birch effect); it occurred with some delay in the larger soil blocks. The effect was not associated with increased humidity, rather with increased availability of oxygen within the soil; therefore we suggest interpreting as reaction of the enzymatic systems of the soil organisms. The respiration rate decreased linearly with decreasing humidity of the soil; RQ dropped below 0.5 in both larger soil blocks for some time; reason is evidently in storage of CO_2 produced in soil water. Rate of respiration registered is also related to consistency of soil; hard “mud pies” respire some 20% less than the small aggregates of sandy character.

Respiration ceased when GWC dropped down to 0.15. RQ was increasing in the second part of the experiments; a few days before the end it reached 1.00 and remained so till the last moment. We think it comes from CO_2 stored in the earlier parts of the drying process which is aerated when amount of water within the soil decreased and air in soil got higher. ΣRQ for the whole budget of the experiments is about 0.7; it is somewhat higher for the smaller soil items (LS) suggesting less hypoxia and may be consistent with partial use of fermentation products like ethanol (RQ = 0.57).

REFERENCES

1. Davidson, E.A., Savage, K., Verchot, L.V. and Navarro, R. (2002) Minimizing Artifacts and Biases in Chamber-Based Measurements

of Soil Respiration. Agricultural and Forest Meteorology, 113, 21-37.

2. Luo, Y. and Zhou, X. (2010) Soil Respiration and the Environment. Academic Press, Burlington.
3. Luan, J., Liu, S., Zhu, X., Wang, J. and Liu, K. (2012). Roles of Biotic and Abiotic Variables in Determining Spatial Variation of Soil Respiration in Secondary Oak and Planted Pine Forests. Soil Biology and Biochemistry, 44, 143-150.
4. Ilstedt, U., Nordgren, A. and Malmer, A. (2000) Optimum Soil Water for Soil Respiration before and after Amendment with Glucose in Humid Tropical Acrisols and a Boreal Mor Layer. Soil Biology & Biochemistry, 32, 1591-1599.
5. Campbell, N.A. (1993) Biology. 3rd Edition, Benjamin/Cummings, Francis Group, Redwood City.
6. Alberts, B. and Johnson, A. (2002) The Molecular Biology of the Cell. Garland, Taylor and Francis Group, New York.
7. McGill, W.B. (2007) The Physiology and Biochemistry of Soil Organisms. In: Paul, E.A., Ed., Soil Microbiology Ecology, and Biochemistry, 3rd Edition, Academic Press, Burlington.
8. Butler, A., Meir, P., Saiz, G., Maracahipes, L., Marimon, B.S. and Grace, J. (2012) Annual Variation in Soil Respiration and Its Component Parts in Two Structurally Contrasting Woody Savannas in Central Brazil. Plant Soil, 352, 129-142.
9. van Suchtelen, H. (1910) Uber die Messung der Lebenstatigkeit der aerobiotischen Bakterienim Boden durch die Kohlensaureproduktion. Centrablattfur Bakteriologie, 28, 46-89. Cited from: Lundegardh, H. (1926) Carbon Dioxide Evolution of Soil and Crop Growth. Soil Science, 23, 417-453.
10. Thomson, B.C., Ostle, N.J., McNamara, N.P., Whiteley, A.S. and Griffith, R.I. (2010) Effects of Sieving, Drying and Rewetting upon Soil Bacterial Community Structure and Respiration Rates. Journal of Microbiological Methods, 83, 69-73.

11. Gaumont-Guay, D.T., Black, T.A., Tim, J., Griffis, T.J., Alan, G., Barr, A.G., Jassal, R.S. and Nesic, Z. (2006) Interpreting the Dependence of Soil Respiration on Soil Temperature and Water Content in a Boreal Aspen Stand. Agricultural and Forest Meteorology, 140, 220-235.

12. Walker, T.S., Harsh, P.B., Grotewold, E. and Vivanco, J.M. (2003) Update on Root Exudation and Rhizosphere Biology. Plant Physiology, 132, 44-51.

13. Chowdhury, N., Burns, R.G. and Marschner, P. (2011) Recovery of Soil Respiration after Drying. Plant Soil, 348, 269-279.

14. Sierra, J. and Renault, P. (1995) Oxygen Consumption by Soil Microorganisms as Affected by Oxygen and Carbon Dioxide Levels. Applied Soil Ecology, 2, 175-184.

15. Jenkins, M.E. and Adams, M.A. (2011) Respiratory Quotients and Q10 of Soil Respiration in Sub-Alpine Australia Reflect Influences of Vegetation Types. Soil Biology & Biochemistry, 43, 1266-1274.

16. Araujo, A.S.F., Cesarz, S., Leite, L.F.C., Borges, C.D., Tsai, S.M. and Eisenhauer, N. (2013) Soil Microbial Properties and Temporal Stability in Degraded and Restored Lands of Northeast Brazil. Soil Biology & Biochemistry, 55, 175-181.

17. Lamy, E., Tran, T.C., Mottelet, S., Pauss, A. and Schoefs, O. (2013) Relationships of Respiratory Quotient to Microbial Biomass and Hydrocarbon Contaminant Degradation during Soil Bioremediation. International Biodeterioration & Biodegradation, 83, 85-91.

18. Blazka, P. and Fischer, Z. (2014) Moisture, Water Holding, Drying and Wetting in Forest Soils. Open Journal of Soil Science, 4, 174-184.

19. Carter, M.R. and Gregorich, E.G., Eds. (2008) Soil Sampling and Methods of Analysis. CRC Press, Taylor & Francis Group, Boca Raton.

20. Grace, C., Hart, M. and Brookes P.C. (2006) Laboratory Manual of the Soil Microbial Biomass Group. Rothamsted Research.

21. Priha, O. and Smolander, A. (2003) Short-Term Uptake of 15NH4 into Soil Microbes and Seedlings of Pine, Spruce and Birch in Potted Soils. Biology and Fertility of Soils, 37, 324-327.

22. Voroney, R.P. (2007) The Soil Habitat. In: Paul, E.A., Ed., Soil Microbiology Ecology and Biochemistry, 3rd Edition, Elsevier, Academic Press, London, 25-49.

23. Angert, A., Yakir, D., Rodeghiero, M., Preisler, Y., Davidson, E.A. and Weiner, T. (2014) Using O2 to Study the Relationships between Soil CO2 Efflux and Soil Respiration. Biogeosciences Discussions, 11, 12039-12068.

24. Birch, H.F. (1964) Mineralisation of Plant Nitrogen Following Alternate Wet and Dry Conditions. Plant Soil, 20, 43-49.

25. Fischer, T. (2009) Substantial Rewetting Phenomena on Soil Respiration Can Be Observed at Low Water Availability. Soil Biology & Biochemistry, 41, 1577-1579.

26. Shi, A. and Marschner, P. (2014) Drying and Rewetting Frequency Influences Cumulative Respiration and Its Distribution over Time in Two Soils with Contrasting Management. Soil Biology & Biochemistry, 72, 172-179.

27. Unger, S., Maguas, C., Pereira, J.S., David, T.S. and Werner, C. (2012) Interpreting Post-Drought Rewetting Effects on Soil and Ecosystem Carbon Dynamics in a Mediterranean Oak Savannah. Agricultural and Forest Meteorology, 154-155, 9-18.

CHAPTER 6

Switchgrass Management Practice Effects on Near-Surface Soil Properties in West-Central Arkansas

Alayna Jacobs[1], Kristofor R. Brye[1], Randy King[2], Joel Douglas[3], Lisa S. Wood[1], Larry C. Purcell[1], Michael Looper[4]

[1]Department of Crop, Soil, and Environmental Sciences, University of Arkansas, Fayetteville, USA

[2]USDA-NRCS Plant Materials Center, Booneville, USA

[3]USDA-NRCS Central National Technology Support Center, Fort Worth, USA

[4]Department of Animal Science, University of Arkansas, Fayetteville, USA

ABSTRACT

Agronomic management practices that maximize monoculture switchgrass (Panicum virgatum L.) yield are generally well understood; however, little is known about corresponding effects of differing switchgrass management practices on near-surface soil properties and processes. The objective of the study was to evaluate the effects of cultivar ("Alamo" and "Cave-in-Rock"), harvest frequency (1- and 2-cuts per year), fertilizer source (poultry litter and commercial fertilizer), and irrigation management (irrigated and non-irrigated) on near-surface soil properties and surface infiltration in a Leadvale silt loam (fine-silty, siliceous, semiactive, thermic, Typic Fragiudult) after four years (2008 through 2011) of consistent management in west-central Arkansas. Irrigating switchgrass increased ($P < 0.01$) soil bulk density in treatment combinations where poultry litter was applied (1.40 $g{\cdot}cm^{-3}$) compared to non-irrigated treatment combinations (1.33 $g{\cdot}cm^{-3}$). Root density was greater ($P = 0.031$) in irrigated (2.62 $kg{\cdot}cm^{-3}$) than in non-irrigated (1.65 $kg{\cdot}cm^{-3}$) treatments when averaged over all other treatment factors. The total infiltration rate under unsaturated conditions was greater ($P = 0.01$) in the 1-cut (33 $mm{\cdot}min^{-1}$) than 2-cut (23 $mm{\cdot}min^{-1}$) harvest treatment combinations when averaged over all other treatment factors, while the total infiltration rate under saturated conditions did not differ among treatment combinations ($P > 0.05$) and averaged 0.79 $mm{\cdot}min^{-1}$. Results from this study indicate that management decisions to maximize switchgrass biomass production affect soil properties over relatively short periods of time, and further research is needed to develop local best management practices to maximize yield while maintaining or improving soil quality.

KEYWORDS

Soil Bulk Density, Infiltration, Extractable Soil Nutrients

1. INTRODUCTION

The United States (US) is the largest consumer of petroleum in the world at an estimated 18.6 million barrels per day [1] . Even though the US is also the third largest producer of petroleum, domestic production cannot keep pace with demand, as 40% of petroleum products consumed in the US are net imports from other countries [1] . Burning fossil fuels also releases carbon dioxide (CO_2). Promoting the generation and development of domestic fuel supplies, while reducing CO_2emissions from fossil fuel combustion, has become a central issue in global policy.

Concerns related to the effects of global warming and possible shortages of finite fossil fuel sources have led to government-mandated regulations encouraging the development of alternative fuels. According to the Energy Policy Act of 2005 and the Renewable Fuel Standard Program of 2007, 36 billion gallons of renewable fuel must be blended into the nation's transportation fuel supply by 2022 [2] . However, only 12% of electricity generated in the US is currently generated from renewable sources, such as wind, hydroelectric, and biomass [3] .

One potential solution is the conversion of plant biomass to fuel. Using plant biomass as a source of cellulosic biofuel as an alternative to fossil fuel is attractive because plants use sunlight for energy and capture existing CO_2 from the atmosphere. Though bioenergy derived from lignocellulosic sources may not provide a complete alternative to the massive energy needs of the US, lignocellulosic bioenergy sources may alleviate a portion of the negative impacts of burning fossil fuels and contribute to an extended supply of fossil fuels for the future [4] .

Plant biomass can be utilized as a fuel in two major ways: by burning or mixing biomass with coal (i.e., co- firing) to generate electricity [5] or conversion of biomass to ethanol. Co-firing is desirable because it reduces the amount of coal burned while maintaining a high level of energy output. Mixing renewable fuels with a non- renewable fuel like coal may also extend the supply of finite resources further into the future. One advantage of using plant biomass rather than corn (Zea mays L.) grain to produce ethanol is the avoidance of sacrificing food for fuel production. Corn is an important food crop for human and livestock consumption. Corn makes up over 95% of the grain rations for cattle, and the US is the world's largest producer of beef [6] . Use of a non-food cellulosic material for conversion to ethanol offers an attractive solution to the problem of competing with the livestock feed industry.

Corn is also an annual crop that must be replanted every year and requires large amounts of fertilizer. The extensive preparation required for conventional corn production is regarded as causing more total soil erosion than any other crop grown in the US [7] . Dedicated energy crops, or crops grown exclusively for conversion to renewable fuel sources, have emerged as potential sources of cellulose. These crops include short-rotation woody crops, such as willow (Salix spp.) and cottonwood (Populus spp.), annual crop residue from small-grain crops and biomass sorghum (Sorghum bicolor [L.] Moench), and perennial grass crops. Ideal energy crops from woody and herbaceous sources have low fertilizer requirements, are relatively cheap and easy to establish, have a perennial growth habit, are easily integrated with existing US farming practices, and are adaptable to sites unsuitable for food or cash crop production.

Switchgrass (Panicum virgatum L.) was recognized by the US Department of Energy in the 1990s as part of the Bioenergy Feedstock Development Program for its ability to produce large quantities of biomass on relatively poor sites. Annual yields of switchgrass in the US average 11.2 $Mg \cdot ha^{-1}$, ranging from 4.5 $Mg \cdot ha^{-1}$ in the northern plains to 23.0 $Mg \cdot ha^{-1}$ in Alabama [8] . Switchgrass is native to all of the US, except in California and the Pacific Northwest, and is adapted to a wide range of soil textures and soil drainage classes [9] . Another desirable aspect of switchgrass production is that the agricultural community is already familiar with perennial grass production, harvesting strategies, and harvesting equipment [8] . In order for producers to commit to growing dedicated energy crops for renewable fuel markets, detailed information about storage, handling, pricing, timing, and transport must be available [10] . Since most of that information does not exist for the majority of US producers, having the option of growing switchgrass as both a source of livestock forage or renewable fuel feedstock may provide added stability for long-term value to producers.

The western portion of Arkansas is largely unsuitable for traditional row-crop production due to steep topography and thin soil compared to the delta region of the Lower Mississippi River basin of eastern Arkansas. However, the average biomass yield for switchgrass grown in Booneville, Arkansas under a single annual harvest regime was 13.4 $Mg \cdot ha^{-1}$ over a four-year period, which was above the national average yield of 11.2 $Mg \cdot ha^{-1}$ estimated by McLaughlin and Kszos [8] . When using the ALMANAC yield prediction computer model, McLaughlin et al. [11] recognized the southeast and south-central US regions as being more suitable than the upper mid-western region for providing dependable future feedstock supplies of switchgrass biomass.

In addition to the projected increase in yield, western Arkansas is widely known for its forage and hay production and for its role in the beef cattle industry. Agricultural operators in western Arkansas are familiar with grass harvest, distribution, and storage and typically own conventional farm equipment that can be used to harvest switchgrass. The cost of producing one dry ton of switchgrass in Arkansas was estimated to be $26.73 in the third year of production, with an expected useful life of 12 years before having to replant [10] . Furthermore, Arkansas is the only US state to be ranked in the top 10 of broiler chicken (Gallus gallusdomesticus), turkey (Meleagris gallopavo), and egg production [12] . As a by-product of poultry production, poultry litter has been widely considered the most valuable manure produced by livestock due to its plant nutrient and organic matter content [13] . Consequently, many producers use dry, surface-applied poultry litter to fertilize hay and forage production pastures.

Past research has identified agronomic procedures for maximizing switchgrass biomass production using varying systems, including cultivar choice, fertilizer application, irrigation, harvest frequency, and row spacing [14] - [17] . Thorough evaluations of the implications associated with switchgrass production systems are needed to seriously determine if switchgrass provides a truly renewable and clean source of domestic energy. Therefore, the objective of this study was to evaluate the effects of switchgrass cultivar (i.e., “Alamo” and “Cave-in-Rock”), water management (i.e., irrigation and no irrigation), fertilizer source (i.e., commercial or poultry litter fertilization), and harvest regime (i.e., once- or twice-harvested per year) after four years of consistent management on near-surface soil properties (i.e., bulk density, soil aggregation, soil pH and electrical conductivity, extractable soil nutrients, and root density) and surface infiltration in west-central Arkansas.

Since “Alamo” previously produced greater biomass yields in all treatment combinations [15] , it was hypothesized that “Alamo” treatment combinations would have lower soil bulk density, greater root density, and greater surface infiltration rate than “Cave-in-Rock” combinations. Treatment combinations fertilized with poultry litter rather than commercial fertilizer were expected to have lower soil bulk density and greater surface infiltration rates as the addition of the organic material associated with the poultry litter would increase porosity. Irrigation was expected to decrease near-surface, mobile soil nutrient levels due to increased water available for leaching. Harvesting twice per year was expected to result in a lower root density, greater bulk density, and lower surface infiltration rates than harvesting once per year. Lastly, it was expected that the greatest-yielding treatment combinations would contain the lowest soil nutrient contents since a greater proportion of plant nutrients would have been removed from the soil in aboveground biomass production.

2. MATERIALS AND METHODS

2.1. Site Description

The US Department of Agriculture's Natural Resource Conservation Service's Plant Materials Center (PMC) was established in Booneville, AR in 1987 and serves the plant material needs of the Southern Ozarks, Arkansas River Valley, Boston, and Ouachita Mountains (Major Land Resource Areas [MLRAs] 116A, 118A, 117, and 119, respectively). The PMC is located along the Petit Jean River in Logan County, AR at an elevation of 146 m [18] and lies along the north edge of the Ouachita National Forest in the eastern portion of the Arkansas Valley and Ridges (MLRA 118A). The 30-yr mean annual precipitation throughout the region is 127 cm and precipitation is delivered more or less evenly throughout the winter (23%), summer (21%), spring (29%), and autumn (27%) seasons [19] . The 30-yr mean annual air temperature for the region is 15.6°C, with a winter minimum of 10.5°C and a summer maximum of 32.4°C [19] .

The study site at the PMC (Figure 1) was located on a Leadvale silt loam (fine-silty, siliceous, semiactive, thermic Typic Fragiudult) [20] . A fragipan is located at a depth of between 30 and 60 cm from the surface such that the Leadvale soil is poorly drained in late winter and early spring due to the presence of a perched water table. The fragipan also causes plants to experience drought stress during the summer by limiting plant rooting depth. The study site was a pasture dominated by tall fescue

(Schedonorus arundinaceus [Schreb] Dumort., nom. Cons.) prior to being prepared for a yield study in 2006 [15] . No fertilizer was applied and the vegetation was kept mowed to a height of approximately 15 cm year-round. Based on soil sampling, soil nutrient levels were brought up to uniform, medium production values prior to planting according to recommended values for phosphorus (P) and potassium (K) for native warm-season grass establishment.

Figure 1. The field study was conducted at the United States Department of Agriculture's Natural Resource Conservation Service's Plant Materials Center (PMC) near Booneville, AR (*).

2.2. Initial Switchgrass Field and Treatment Establishment

"Alamo" and "Cave-in-Rock" switchgrass cultivars were planted in 12.2 by 12.2 m plots on 5 March 2007. Each plot contained one switchgrass cultivar and fertilizer source. Plots were seeded at a rate of 4.4 $kg \cdot ha^{-1}$ pure live seed (457 $seed \cdot m^{-2}$) and planted with a no-tillage native grass drill (Sukup 2050 series, Jonesboro, AR). After drilling, the seedbed was rolled with a water-filled roller to establish good seed-to-soil contact. Temporary sprinkler irrigation was applied to all plots for initial seed emergence and establishment. A permanent sprinkler irrigation system was installed in replicated irrigation treatments in summer 2007. Each 12.2 by 12.2 m plot was divided in half, with one subplot harvested once per year and the other subplot harvested twice per year. There were a total of 48 plots encompassing three replications of each cultivar-irrigation-fertilizer source-harvest frequency treatment combination.

Rain gauges were placed in the irrigated plots to calibrate the irrigation delivery system. Irrigated plots received 2.54 cm of irrigation water from a nearby surface pond per week during June through August from 2008 to 2011. The average annual precipitation at the study site was 131 cm [18] , which was slightly greater than the 30-yr average of 121 cm for the region [19] . Annual rainfall represented 127% in 2008, 124% in 2009, 70% in 2010, and 110% in 2011 of the 30-yr average rainfall [18] [19] . Average annual air temperature represented 90% in 2008, 89% in 2009, 93% in 2010, and 99% in 2011 of the 30-yr average [18] [19] .

The study site was burned each year in early March to remove residue stubble, stimulate switchgrass seed production for wildlife, remove surface residue for native pollinator nesting habitat, and to create corridors for other ground-nesting wildlife species [21] . Fertilizer treatments were applied when green foliage appeared in early April each year (i.e., 2008 to 2011). Poultry litter (PL) was applied at a rate of 4.5 $Mg \cdot ha^{-1}$, while commercial fertilizer (CF) was applied to match nitrogen (N), P, and K levels in the applied PL. Litter batches were analyzed annually for nutrient concentrations prior to application.

Two harvest frequencies were imposed to test their effects on annual aboveground biomass production. A single harvest was made in November after the first killing frost for the 1-cut system. In the 2-cut system, harvests occurred twice per year. The first harvest occurred in June just prior to the boot stage when the seed head emerges and the second harvest occurred after a killing frost in November. Additional switchgrass harvesting details and results of the initial yield study were reported in Jacobs and King [15].

2.3. Soil, Plant, and Infiltration Measurements

In July 2012, soil samples were collected in all plots from the 0- to 10- and 10- to 20-cm depth intervals for bulk density, extractable soil nutrients, and soil particle-size analyses. In the top 10 cm, bulk density samples were collected manually with a 5-cm outside diameter, stainless steel core chamber and a slide hammer. For the 10- to 20-cm depth, bulk density samples were collected with a 5-cm diameter, stainless steel mechanical hydraulic probe. Samples from both depths were dried in a forced-air dryer at 70°C for 48 hours, and then weighed for bulk density determinations. Soil from the bulk density samples was sieved through a 2-mm mesh screen and used to measure particle-size distribution using a modified 12-hour hydrometer method [22] . Mehlich-3 extractable soil nutrients (i.e., P, K, Ca, Mg, S, Na, Fe, Mn, Zn, and Cu) were determined in a 1:2 soil-water extract [23] [24] . Soil pH and electrical conductivity (EC) potentiometrically in a 1:2 soil-water paste.

In August 2013, soil samples were collected from the 0- to 5- and 5- to 10-cm depths using a 4.8-cm diameter core chamber and slide hammer to measure aggregate stability (AS) (i.e., water-stable aggregates > 0.25-mm diameter) using a wet-sieving procedure [25] . Soil core samples were collected from areas between switchgrass crowns in the switchgrass rows. Two replications were collected from each plot, mixed together for one sample per depth interval, and allowed to air dry for 15 days. Approximately 400 g of air-dried soil were used for the wet-sieving procedure, where soil was plunged in tap water at 30 cycles per minute for 5 minutes. Soil retained on the mesh openings of each of five sieve sizes (i.e., 4, 2, 1, 0.5, and 0.25 mm) was rinsed into an aluminum container with tap water. Water was decanted from the container, and the remaining soil was weighed after being dried at 70°C for 24 hours. The soil mass from each container was summed to calculate the total mass of water- stable aggregates from each plot.

One soil sample for root density was collected per plot in September 2013 from the 0- to 15-cm depth interval using a 7.3-cm diameter core chamber and slide hammer, and prepared according to the procedures followed by Brye and Riley [26] . Root samples were washed on a 2-mm mesh screen to collect the root material, dried at 55°C for 24 hours, and weighed. Root density samples were collected from areas between switchgrass crowns in the switchgrass rows.

In November 2013, double-ring infiltrometers were used to measure surface infiltration rates two days after a soaking rainfall. Mature switchgrass was trimmed using hedge trimmers and residue carefully removed prior to placement of infiltrometers. One double-ring infiltrometer measurement was conducted in each cultivar-harvest frequency-fertilizer source treatment combination. In order to maximize potential infiltration differences among treatment combinations, infiltration measurements were only conducted in the non-irrigated treatment. Double- ring infiltrometers were placed between switchgrass crowns in the switchgrass rows after switchgrass was mowed to a height of 6 cm. The outer-ring diameter was 30 cm, the inner-ring diameter was 6 cm, and rings were 10 cm in height. The infiltrometer was inserted approximately 2 cm deep into the soil and the outer ring was filled with tap water to act as a buffer between dry soil outside the outer ring and saturated soil inside the inner ring. The inner ring was filled with tap water and the distance from the top of the soil to the water surface in the inner ring was measured at 0, 1, 2, 3, 4, 5, 8, 10, 12, 15, 18, and 20 minutes.

Double-ring infiltration measurements were paired with mini-disk tension infiltrometer (Decagon, Pullman, Washington) measurements, where infiltrometer tension was set at −2 cm for each measurement. Two mini-disk infiltrometer measurements were conducted in each plot, with one measurement collected from the center-ring area immediately following double-ring infiltrometer measurements and one measurement collected in a nearby location. The infiltrative surface of the mini-disk infiltrometer was 4.5 cm. The two mini-disk measurements were performed between switchgrass crowns in the switchgrass rows in each plot for a total of 48 measurements. The height of the water inside the 2.1-cm diameter mini-disk infiltrometer water reservoir was recorded at 0, 1, 2, 3, 4, 5, 8, 10, 12, 15, 18, and 20 minutes. The total surface infiltration rate was calculated for both sets of infiltration measurements by dividing the total amount of infiltrated water by 20 minutes.

2.4. Statistical Analyses

The effects of cultivar, harvest frequency, irrigation, fertilizer source, soil depth, and their interactions on soil bulk density, soil chemical properties, particle-size distributions, and AS were evaluated by analysis of variance (ANOVA) using PROC GLM in SAS (version 9.3, SAS Institute, Inc., Cary, NC). A separate ANOVA was conducted to evaluate the effects of cultivar, harvest frequency, irrigation, fertilizer source, and their interactions on root density using SAS. A separate ANOVA was also conducted to evaluate the effects of cultivar, harvest frequency, fertilizer source, and their interactions on total surface infiltration rate for both sets of infiltration data. All treatments were treated as fixed effects for all ANOVAs conducted. An analysis of covariance (ANCOVA) was conducted to evaluate the effects of cultivar, harvest frequency, and fertilizer source on the relationship between the natural logarithm of the infiltration rate and the natural logarithm of the mid-point of the measurement time interval. When appropriate, means were separated by least significant difference (LSD) at $\alpha = 0.05$.

3. RESULTS

3.1. Soil Properties

3.1.1. Particle-Size Distribution

Throughout the study site, sand, silt, and clay in the top 20 cm varied somewhat. Though sand, silt, and clay distributions remained within the range of a silt-loam texture in all treatment combinations, sand, silt, and clay contents in the top 20 cm differed slightly among various treatment combinations ($P < 0.05$), with the largest differences occurring between soil depths, which was expected. In both depth intervals, sand content was greatest in the irrigated-“Alamo” treatment combinations (Table 1). Sand content was lowest in the top 10 cm in the non-irrigated-“Cave-in-Rock” (26.1%) and in the 10 - 20 cm depth interval in the non-irrigated-“Alamo” (27.5%) treatment combinations, which did not differ (Table 1). Averaged over all other treatment factors, silt content was greater ($P < 0.01$) in the top 10 cm than in the 10- to 20-cm depth interval (50.4 and 48.2%, respectively). Silt content was greater ($P = 0.04$) in the PL (50%) than in the CF (48.6%) treatment combinations. Clay content was greater ($P = 0.03$) in the CF (21.7%) than in the PL treatment combinations (20.6%). Silt content was greater ($P < 0.01$) in “Cave-in-Rock” (50.2%) than in “Alamo” (48.4%) treatments. Despite the significant fertilizer source effect on silt and clay and significant cultivar effect on silt content, all silt and clay differences between treatments were less than 2%. Small differences of this magnitude are not likely to affect agronomic management or productivity.

Averaged over harvest frequency, fertilizer source, and irrigation, clay content also differed between cultivars within soil depths ($P < 0.01$). As expected, clay content increased with increasing depth from the top 10 cm (19.5%) to the 10- to 20-cm depth interval (22.9%) for both cultivars (Figure 2). Clay content was greatest and least for “Alamo” treatment combinations in the 10- to 20-cm (22%) and in the 0- to 10-cm (17%) depth intervals, respectively (Figure 2).

The observed differences in sand, silt, and clay contents between soil depths were expected based on the reported textural classes of the top two horizons of the Leadvale silt-loam official series description [20] . Clay accumulation with increasing depth was expected, as the surface Ap horizon grades to an argillic Bt horizon at approximately 20 cm [20] . Despite some relatively minor differences in particle-size distribution among various treatment combinations, differences were not large enough to change the soil taxonomic classification anywhere throughout the study site and would likely not influence practical agronomic management decisions related to switchgrass production.

Table 1. Summary of the effects of soil depth, cultivar, and irrigation on sand and extractable soil manganese (Mn) contents and electrical conductivity (EC).

Soil property	Soil depth (cm)	Alamo		Cave-in-Rock	
		Irrigated	Non-irrigated	Irrigated	Non-irrigated
Sand (%)	0 - 10	32.6 a*	31.4 ab	30.3 abc	26.1 d
	10 - 20	30.3 ab	27.5 cd	29.0 bc	28.9 bc
Mn ($kg \cdot ha^{-1}$)	0 - 10	183.2 b*	170.9 b	179.3 b	169.4 b
	10 - 20	230.7 a	168.9 b	178.7 b	185.5 b
EC ($dS \cdot m^{-1}$)	0 - 10	0.070 a*	0.072 a	0.071 a	0.070 a
	10 - 20	0.060 bc	0.053 c	0.053 c	0.067 ab

*For each soil property, means followed by different letters differ significantly at $P < 0.05$.

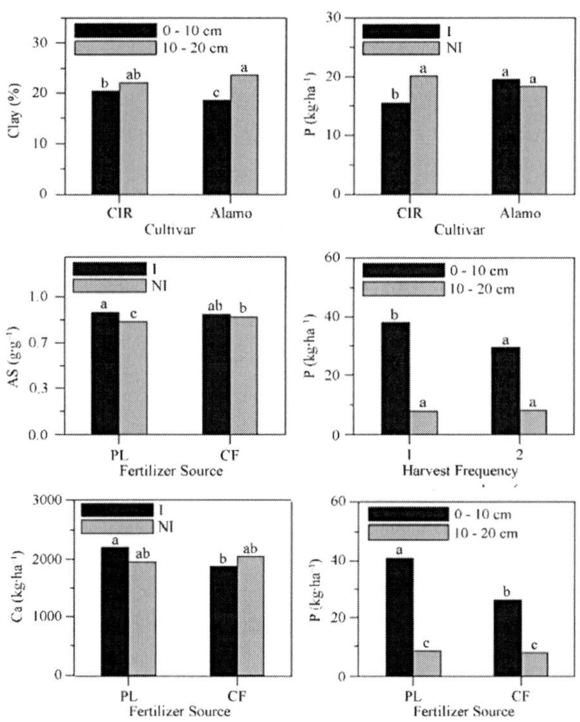

Figure 2. Soil clay content, aggregate stability (AS), and extractable soil phosphorus (P) and calcium (Ca) as affected by soil depth, irrigation [irrigated (I) and non-irrigated (NI)], cultivar ["Cave-in-Rock" (CIR) and "Alamo"], harvest frequency [one (1) harvest per year and two (2) harvests per year], and/or fertilizer source [poultry litter (PL) and commercial fertilizer (CF)]. Different letters atop bars within a soil property indicate a significant difference at P < 0.05.

3.1.2. Bulk Density

Bulk density was affected by several experimental treatment factors. As expected, averaged over all other treatment factors, soil bulk density was greater ($P < 0.001$) in the 10- to 20-cm depth (1.44 $g \cdot cm^{-3}$) than in the top 10 cm (1.30 $g \cdot cm^{-3}$). Averaged over soil depths and cultivars, soil bulk density was also greater ($P = 0.004$) in the irrigated-PL treatment combination with either harvest frequency (1.40 $g \cdot cm^{-3}$) than the non-irrigated-1-cut treatment combination with either fertilizer source (1.33 $g \cdot cm^{-3}$). Soil bulk density was unaffected ($P > 0.05$) by switchgrass cultivar.

3.1.3. Aggregate Stability

Similar to bulk density, aggregate stability was affected by several experimental treatment factors. Averaged over soil depth, cultivar, and harvest frequency, AS was greater ($P = 0.03$) under irrigated (0.93 $g \cdot g^{-1}$) than non-irrigated (0.86 $g \cdot g^{-1}$) treatments with PL, but did not differ between irrigation treatments with CF (0.86 $g \cdot g^{-1}$; Figure 2). Averaged over irrigation, the greatest ($P = 0.02$) AS was measured in the "Alamo"-CF-2-cut treatment combination in the 10- to 20-cm depth (0.93 $g \cdot g^{-1}$), while the lowest AS was measured in the "Cave- in-Rock"-PL-1-cut treatment combination in the top 10 cm depth (0.87 $g \cdot g^{-1}$; Table 2).

3.1.4. Soil pH and Electrical Conductivity

Soil pH was unaffected by switchgrass cultivar, irrigation, harvest frequency, or soil depth ($P > 0.05$), but differed between fertilizer sources ($P = 0.001$). As expected, averaged over all other treatment factors, soil pH was greater in the PL (pH = 6.1) than in the CF (pH = 5.9) treatment. Though soil pH differed between fertilizer sources by only 0.02 units, nutrient availability and buffering capacity were likely different as well between fertilizer sources.

Similar to bulk density and aggregate stability, soil EC was affected by several experimental treatment factors. Averaged over fertilizer source and soil depth, soil EC was greatest ($P = 0.016$) in "Cave-in-Rock" (0.070 $dS \cdot m^{-1}$) and least in "Alamo" (0.056 $dS \cdot m^{-1}$) treatments without irrigation in the 2-cut harvest frequency. Averaged over harvest frequency and fertilizer source, soil EC was generally greater ($P = 0.006$) in the top 10 cm than in the 10- to 20-cm depth and was the lowest in the non-irrigated-"Alamo" and irrigated-"Cave-in-Rock" treatment combinations in the 10- to 20-cm depth interval, which did not differ (Table 1). Averaged over cultivar and irrigation, soil EC was also greatest ($P = 0.020$) in the 1-cut-PL treatment combination in the top 10 cm (0.081 $dS \cdot m^{-1}$) compared to all other treatment combinations (Table 3). Soil EC did not differ among harvest- frequency-fertilizer-source treatment combinations in the 10- to 20-cm depth and averaged 0.058 $dS \cdot m^{-1}$ (Table 3).

3.1.5. Extractable Soil Nutrients

Similar to bulk density, aggregate stability, and soil EC, numerous extractable soil nutrients were affected by multiple experimental treatment factors. Averaged over harvest frequency and fertilizer source, extractable soil Mn was greater ($P = 0.026$) in the irrigated-"Alamo" treatment combination in the 10- to 20-cm depth interval (230.2 $kg \cdot ha^{-1}$) than in all other treatment combinations, which did not differ and averaged 176.6 $kg \cdot ha^{-1}$ (Table 1). Unexpectedly, averaged over all other factors, "Cave-in-Rock" treatments had greater ($P = 0.033$) extractable soil K (95 kg K ha^{-1}) and lower ($P = 0.019$) extractable soil Na (43 kg Na ha^{-1}) compared to "Alamo" treatments (79 kg K and 64 kg Na ha^{-1}, respectively). Averaged over all other treatment factors, extractable soil Ca was greater ($P = 0.031$) in the 10- to 20-cm depth (2079 $kg \cdot ha^{-1}$) than in the top 10 cm (1965 kg Ca ha^{-1}). In contrast, extractable soil Fe was greater ($P < 0.001$) in the top 10 cm (216 kg Fe ha^{-1}) than in the 10- to 20-cm depth (164 kg Fe ha^{-1}).

Averaged over irrigation, cultivar, and harvest frequency, extractable soil P, K, Mg, S, Na, and Zn contents followed similar trends related to treatment combinations and were generally greater ($P < 0.04$) in the top 10 cm compared to the 10- to 20-cm depth. Extractable soil P (Figure 2), S, Zn (Figure 3), K, and Mg (Figure 4) contents were greatest in the PL treatment combinations in the top 10 cm and the lowest in the 10- to 20-cm depth regardless of fertilizer source, which did not differ. In contrast, extractable soil Na was generally greater in the 10- to 20-cm depth compared to the top 10 cm and was greatest in the 10- to 20-cm-depth-PL and lowest in the 0- to 10-cm CF treatment combination (Figure 3).

Table 2. Summary of the effects of soil depth, harvest frequency, cultivar, and fertilizer source on aggregate stability (AS).

Soil property	Soil depth (cm)	Harvest frequency	Alamo		Cave-in-Rock	
			PL†	CF†	PL	CF
AS ($g \cdot g^{-1}$)	0 - 10	1	0.90 bcde*	0.91 abc	0.87 f	0.89 cdef
		2	0.89 cdef	0.92 ab	0.91 abc	0.88 def
	10 - 20	1	0.90 bcde	0.92 ab	0.88 ef	0.89 cdef
		2	0.92 ab	0.93 a	0.90 bcde	0.90 bcd

†PL, poultry litter; CF, commercial fertilizer. *Means followed by different letters differ significantly at $P < 0.05$.

Table 3. Summary of the effects of harvest frequency, soil depth, and fertilizer source on extractable soil copper (Cu) content and electrical conductivity (EC).

Soil property	Harvest frequency	0 - 10 cm depth		10 - 20 cm depth	
		PL†	CF†	PL	CF
Cu ($kg \cdot ha^{-1}$)	1	2.9 a*	2.0 c	1.0 d	0.9 de
	2	2.5 b	2.01 c	1.0 de	0.8 e
EC ($dS \cdot m^{-1}$)	1	0.081 a*	0.067 b	0.059 c	0.059 c
	2	0.068 b	0.067 b	0.060 bc	0.055 c

†PL, poultry litter; CF, commercial fertilizer. *For each soil property, means followed by different letters differ significantly at $P < 0.05$.

Averaged over irrigation, cultivar, and fertilizer source, extractable soil P, K, Mg, and Zn contents also differed between soil depths within harvest frequency. Extractable soil P (Figure 2), Zn (Figure 3), K, and Mg (Figure 4) contents followed similar trends related to treatment combinations and were generally greater (P < 0.02) in the top 10 cm compared to the 10- to 20-cm depth. Extractable soil P (Figure 2), Zn (Figure 3), K, and Mg (Figure 4) contents were greatest in the 1-cut treatment combination in the top 10 cm, which supported the hypothesis that harvesting once after switchgrass senescence rather than twice per year allows greater retention of extractable soil nutrients. Extractable soil P (Figure 2), Zn (Figure 3), K, and Mg (Figure 4) contents were lower in the 10- to 20-cm, which did not differ between harvest frequencies, than in the 0- to 10-cm depth.

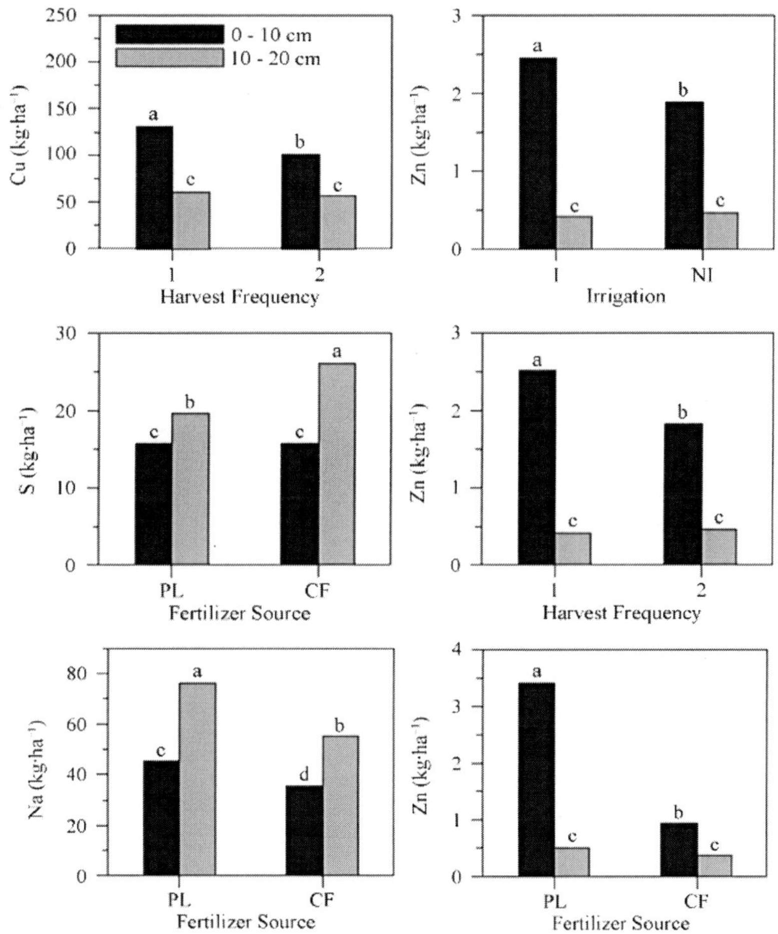

Figure 3. Extractable soil copper (Cu), zinc (Zn), sulfur (S), and sodium (Na) contents as affected by soil depth, irrigation [irrigated (I) and non-irrigated (NI)], harvest frequency [one (1) harvest per year and two (2) harvests per year], and/ or fertilizer source [poultry litter (PL) and commercial fertilizer (CF)]. Different letters atop bars within a soil property indicate a significant difference at $P < 0.05$.

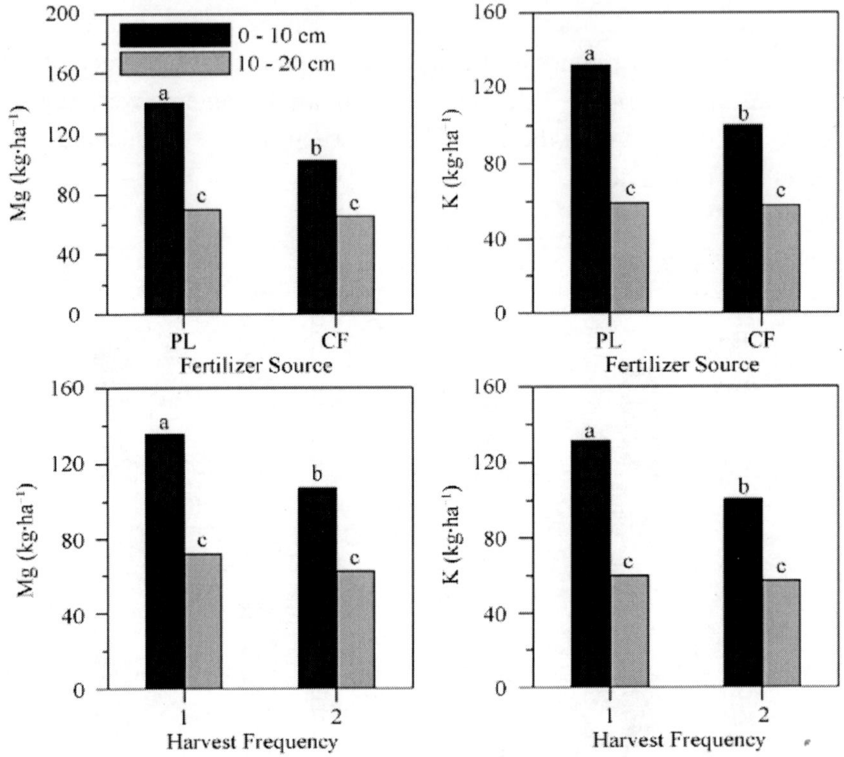

Figure 4. Extractable soil potassium (K) and magnesium (Mg) as affected by soil depth and fertilizer source [poultry litter (PL) and commercial fertilizer (CF)] or harvest frequency [one (1) harvest per year and two (2) harvests per year]. Different letters atop bars within a soil property indicate a significant difference at $P < 0.05$.

In addition, averaged over cultivar, harvest frequency, and soil depth, extractable soil Ca (Figure 2) and Cu (Figure 3) contents followed similar trends, as both were the lowest ($P < 0.03$) in the irrigated-CF and the greatest in the irrigated-PL treatment combination. Averaged over harvest frequency, fertilizer source, and soil depth, extractable soil P content was lower ($P = 0.016$) in the irrigated-"Cave-in-Rock" than in the "Alamo" of either irrigation treatment and in the non-irrigated-"Cave-in-Rock" treatment combination, which did not differ (Figure 2).

Averaged over cultivar, harvest frequency, and fertilizer source, extractable soil Zn content differed between irrigation treatments within soil depths. While the greatest (P = 0.025) extractable Zn content was in the irrigated-0- to 10-cm depth compared to all other treatment combinations, extractable Zn content was generally greater in the top 10 cm than in the 10- to 20-cm depth (Figure 3). Extractable soil Zn was lowest in the 10- to 20-cm depth of either irrigation treatment compared to all other treatment combinations (Figure 3). Averaged over soil depth, the greatest (P = 0.043) extractable Zn content was in the irrigated-"Alamo"-PL (2.3 kg·ha^{-1}) treatment, but did not differ from the "Cave-in-Rock"-PL treatment of either irrigation combination. The lowest extractable Zn content was in the irrigated-"Alamo"-CF (0.6 kg·ha^{-1}) treatment, but did not differ from all other CF treatments in combination with either irrigation or cultivar.

Averaged over irrigation and soil depth, extractable soil Cu (P = 0.009) was greatest in the "Alamo"-PL-1-cut and in the "Cave-in-Rock"-PL treatment combinations under both harvest frequencies (2.0 kg Cu ha^{-1}) compared to all other treatment combinations. Averaged over cultivar and irrigation, extractable soil Cu content was greater (P = 0.038) in the 1-cut-PL treatment combination in the top 10 cm (2.9 kg Cu ha^{-1}) than in all other treatment combinations, which were <2.0 kg Cu ha^{-1} (Table 4). Extractable soil Cu content was generally greater in the top 10 cm and in treatments that included PL. Averaged over harvest frequency and irrigation, extractable soil Cu content was greatest (P = 0.032) in the "Cave-in-Rock"-PL treatment combination in the top 10 cm (2.9 kg Cu ha^{-1}) than all treatment combinations in the 10- to 20-cm depth, which were all <1.1 kg Cu ha^{-1} and did not differ.

Table 4. Summary of the effects of harvest frequency, cultivar, and fertilizer source on extractable soil copper (Cu) and iron (Fe) contents.

Soil property	Harvest frequency	Alamo		Cave-in-Rock	
		PL†	CF†	PL	CF
Cu (kg·ha^{-1})	1	2.0 a*	1.5 b	1.9 a	1.5 b
	2	1.6 b	1.4 b	2.0 a	1.4 b
Fe (kg·ha^{-1})	1	212.5 a*	184.2 b	193.3 ab	189.5 b
	2	195.5 ab	188.8 b	192.6 ab	160.9 c

†PL, poultry litter; CF, commercial fertilizer. *For each soil property, means followed by different letters differ significantly at $P < 0.05$.

3.2. Root Density

In contrast to the initial hypothesis that irrigation would have minimal effect on belowground switchgrass growth in the climatic region of west-central Arkansas, averaged over all other treatment factors, switchgrass root density was greater ($P = 0.031$) in the irrigated (2.62 $kg{\cdot}m^{-3}$) than in the non-irrigated (1.65 $kg{\cdot}m^{-3}$) treatment.

It was also hypothesized that increasing the harvest frequency from 1 to 2 cuts per year would significantly decrease switchgrass root density. However, results did not support this initial hypothesis as root density was also unaffected ($P > 0.05$) by switchgrass cultivar and fertilizer source.

3.3. Infiltration Characteristics

3.3.1. Total Infiltration

Total water infiltration rates measured with the double-ring infiltrometer did not differ ($P > 0.05$) among any treatment combination. Total infiltration rates ranged from a low of 0.2 $mm{\cdot}min^{-1}$ in the non-irrigated-"Alamo" treatment combination with either fertilizer source to a high of 2.4 $mm{\cdot}min^{-1}$ in the non-irrigated-"Cave-in- Rock"-PL treatment combination and averaged 0.79 $mm{\cdot}min^{-1}$ across all treatment combinations.

Unlike total infiltration rates measured with the double-ring infiltrometer, total infiltration rates measured with the mini-disk infiltrometer at a tension of −2 cm were greater ($P = 0.034$) in the 1-cut (33 $mm{\cdot}min^{-1}$) than in the 2-cut (24 $mm{\cdot}min^{-1}$) treatment. The total tension infiltration rate was unaffected ($P > 0.05$) by cultivar, irrigation, or fertilizer source.

3.3.2. Relationship between Infiltration Rate and Time

1) Double-ring infiltrometer

Though total infiltration measured with the double-ring infiltrometer did not differ among experimental treatments, averaged over fertilizer sources, the slope of the linear regression characterizing the natural logarithm of the infiltration rate against the natural logarithm of the mid-point time was greatest (P = 0.004) for the "Cave-in-Rock"-2-cut and "Alamo"-1-cut treatment combinations, which did not differ (Figure 5). The slope was smallest for the "Cave-in-Rock"-1-cut and "Alamo"-2-cut treatment combinations, which did not differ (Figure 5). In addition, averaged over cultivar, the slope of the linear regression characterizing the natural logarithm of the infiltration rate against the natural logarithm of the mid-point time was greater (P = 0.048) for the CF-1-cut than the PL-1-cut treatment combinations (Figure 5).

2) Mini-disk infiltrometer

Unlike the double-ring infiltrometer, the slope of the linear regression equations characterizing the natural logarithm of the infiltration rate against the natural logarithm of the mid-point time was positive and numerically greatest (P = 0.026) for the "Alamo"-CF-2-cut and numerically smallest and negative for the "Alamo"-PL-2-cut and "Cave-in-Rock"-CF-2-cut treatment combinations, which did not differ (Table 5). The y-intercept of the linear relationship was positive and numerically greatest (P = 0.049) in the "Cave-in-Rock"-PL-1-cut and positive and numerically smallest in the "Alamo"-CF-2-cut treatment combination (Table 5).

Table 5. Summary of parameters that characterize the relationship between the natural logarithm of the mini-disk (tension) infiltration rate and time.

Cultivar	Fertilizer source	Harvest frequency	Slope (m)	Y-intercept
Alamo	Commercial fertilizer	1	−0.088 ab*	1.109 b
		2	0.047 a	0.253 c
	Poultry litter	1	−0.126 abcd	1.270 ab
		2	−0.212 cd	1.266 ab
Cave-in-Rock	Commercial fertilizer	1	0.131 abcd	1.190 ab
		2	−0.196 bcd	0.873 bc
	Poultry litter	1	−0.126 abcd	1.852 a
		2	−0.106 abc	0.909 bc

*Means followed by different letters differ significantly at $P < 0.05$.

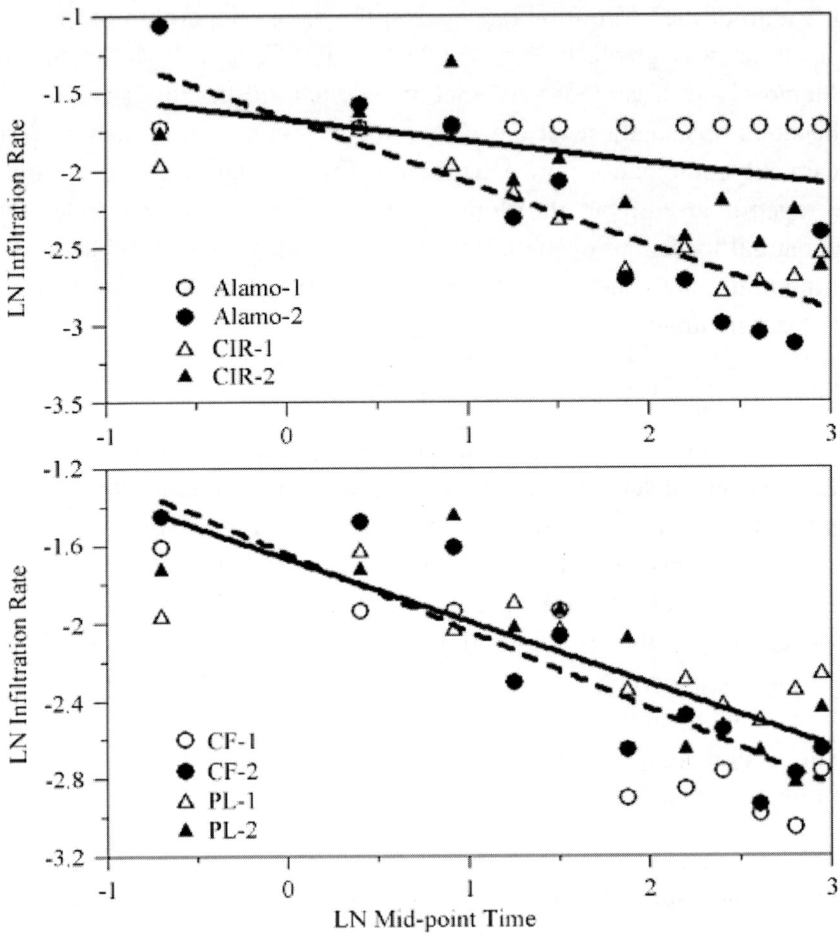

Figure 5. Natural logarithm (LN) of the double-ring infiltration rate per LN mid-point time as affected by cultivar ["Cave-in-Rock" (CIR) and "Alamo"] and harvest frequency [one (1) harvest per year and two (2) harvests per year] (top panel), or fertilizer source [Poultry Litter (PL) and Commercial Fertilizer (CF)] and harvest frequency (bottom panel). Solid regression line (top) includes CIR-2 and Alamo-1, for which slopes did not differ. Dashed regression line (top) includes CIR-1 and Alamo-2, for which slopes did not differ. Solid regression line (bottom) includes PL-1, CF-2, and PL-2, for which slopes did not differ. Dashed regression line (bottom) includes CF-1, PL-2, and CF-2, for which slopes did not differ.

4. DISCUSSION

Overall, results of this study indicated treatments imposed on the switchgrass production system affected near- surface properties in the top 20 cm and surficial processes. However, treatment combinations that produced significant differences in the amount of aboveground biomass production noted previously by Jacobs and King [15] did not necessarily produce significant differences in the measured soil properties in this study.

4.1. Cultivar Effects

Switchgrass cultivar was not expected to affect most soil properties; however, AS and silt concentration, and Na, Fe, P, and K contents differed between cultivars when averaged over all other treatment factors. In addition, clay concentration differed by depth within cultivars when averaged over irrigation, harvest frequency, and fertilizer source. While the silt and clay differences were likely residual inherent differences from before the study was initiated, and assuming similar nutrient concentrations between cultivars, greater soil extractable P and K contents in "Cave-in-Rock" treatments may have resulted from lower nutrient removal rates according to lower biomass production compared to "Alamo" treatments [15] . Average biomass yields at this site over a four-year period (i.e., 2008 through 2011) were significantly greater from "Alamo" than from "Cave-in-Rock" treatments [15] ; thus, greater plant uptake and removal of macronutrients resulting in lower extractable soil P and K contents would be expected.

While extractable soil nutrient differences were relatively straightforward, AS depends on many factors related to soil, water, and plant relationships. Previous research in Arkansas has shown that aggregate stability in the top 10 cm is greater in undisturbed native prairie or prairie restorations than in cultivated, row-crop agroecosystems. Aggregate stability in row-crop production varied from 0.05 $g \cdot g^{-1}$ in a non-irrigated, wheat (Triticum aestivum)-soybean (Glycine max), double-cropped system managed consistently for 10 years [27] to 0.11 $g \cdot g^{-1}$ in irrigated and non-irrigated rice (Oryza sativa)-

wheat double-cropped systems managed consistently for 6 years [28] . In contrast, Brye and Riley [26] reported total water-stable aggregate concentrations of 0.44 $g \cdot g^{-1}$ in an undisturbed native prairie in northwest Arkansas. Interestingly, AS measured in this study, which averaged 0.89 $g \cdot g^{-1}$ in the top 10 cm (Table 2), were two times greater than those measured in a native prairie ecosystem (0.44 $g \cdot g^{-1}$) [26] . Differences may be attributed to the intensive nature of treatment combinations imposed in this study to maximize aboveground biomass production, such as irrigation and fertilizer applications. Greater AS observed in this study may also be attributed to complex interactions involving plant roots and soil aggregation and show the similarity of perennial, monoculture switchgrass production to native grasslands.

Differences in root density could influence AS, but results from this study showed that root density did not differ between cultivars. Soil organic matter (SOM) may also influence AS, but SOM was not measured in this study. The 0- to 15-cm depth interval where root density was measured in this study may not have captured enough of each cultivar's unique rooting characteristics, which may explain why AS and not root density differed between cultivars when averaged over all other factors. Frank et al. [29] reported that 50% of switchgrass root mass occurred in the top 30 cm of soil. In addition, Jung et al. [30] reported that switchgrass root partitioning between fine and coarse roots is dynamic and may be altered with differing levels of soil fertility. Due to differing extractable soil nutrient contents between cultivars, the partitioning between fine and coarse roots may not have been accurately captured for all treatments by the 2-mm sieve size used for the root washing procedure.

4.2. Irrigation Effects

Similar to cultivar effects, it was hypothesized that irrigation treatments would have little influence on soil properties. However, irrigation treatments produced significant differences in root density. Though application of irrigation to the switchgrass crop did not significantly increase biomass yields [15] , irrigation increased root density compared to the non-irrigated treatment when averaged over all other treatment factors. It was clear that irrigation directly stimulated root growth and, as a result, indirectly altered additional soil properties measured in this study, such as AS.

Averaged over all other treatment factors, AS did not differ between irrigation treatments. However, increased AS been linked to greater root density, as roots encourage the formation and stabilization of soil aggregates [31] . The lack of a significant irrigation effect on AS in this study was in contrast to Singer et al. [32] , where the rewetting of dry soil during irrigation cycles decreased the concentration of soil macroaggregates compared to a non-irrigated site. The destruction of soil aggregates due to irrigation was attributed to an increase in air pressure inside dry soil macroaggregates during the rewetting process (i.e., slaking) [32] . Soil aggregates may not have experienced sufficient wetting and drying for slaking to occur in west-central Arkansas, given irrigation did not significantly increase switchgrass biomass yield [15] .

Whalen and Chung [33] compared soil aggregate stability with differing application rates of cattle manure on cropland in Canada. While Whalen and Chung [33] hypothesized that soil macroaggregate (>12.1 mm) stability would increase with increased manure application rate, results showed that the proportion of soil macroaggregates actually decreased with manure application compared to no manure application. The dispersive effects of monovalent cations and foreign soil present in cattle manure were cited as the explanation for their results [33] . Edaño [34] also determined that the application of manure fertilizer (i.e., cattle manure or PL) compared to CF did not encourage macroaggregate formation in switchgrass plots harvested in both 1- and 2-cut systems.

The dispersive effects of monovalent cations typically present in PL, such as Na, may have affected soil properties evaluated in this study. For example, irrigation of plots fertilized with PL may have caused greater dispersion by further mobilizing monovalent cations. In this study, extractable soil Na was greater in the 10- to 20-cm depth than in the top 10 cm. Results from this study also further support the conclusions of Whalen and Chung [33] and Edaño [34] as soil bulk density was greater in the irrigated-PL treatment combination with either harvest frequency (1.41 $g \cdot cm^{-3}$) than in the non-irrigated-1-cut treatment combination with either fertilizer source (1.33 $g \cdot cm^{-3}$).

4.3. Fertilizer Source Effects

The application of two different fertilizer sources over a four-year period [15] produced significant differences in extractable soil nutrient levels. Even though significant differences between fertilizer sources were present among extractable soil nutrient levels after the four-year study period, biomass yield was unaffected by fertilizer source [15] . In contrast, Edaño [34] reported that switchgrass biomass yield was greater with PL compared to CF application, switchgrass seeded with legumes, and an unfertilized control.

Poultry litter was applied at a rate of 4.5 $Mg{\cdot}ha^{-1}$ and corresponding amounts of N, P, and K were applied as CF. Poultry litter contains more than just N, P, and K, and significant differences between the levels of extractable soil nutrients were apparent in this study. Only one soil nutrient tested (i.e., Mn) did not differ between fertilizer sources when averaged across all other treatment factors. Soil collected from treatment combinations fertilized with PL contained significantly greater contents of extractable soil P, K, Ca, Mg, S, Na, Fe, Zn, and Cu than CF treatment combinations. Edaño [34] observed similar increases in extractable soil P when switchgrass was fertilized with cattle manure and PL instead of CF.

In addition, both soil pH and EC were significantly greater in PL than in CF treatment combinations, which were expected. Results from this study regarding soil pH changes in “Alamo” switchgrass from fertilizer sources were similar to those reported by Edaño [34] in Oklahoma. Soil pH slightly increased when fertilized with cattle manure or PL and slightly decreased when fertilized with CF compared to an unfertilized control [34] . The differences in pH would not likely affect agronomic management decisions, especially given that soil pH values of 5.5 to 6.6 are considered optimum for native, warm-season grass establishment in Arkansas [21] .

4.4. Harvest Frequency Effects

The previous yield study by Jacobs and King [15] largely focused on the effects of harvest frequency on biomass yield due to the dual-purpose nature of switchgrass production as a potential livestock forage and as a potential renewable energy feedstock. Harvesting once per year in the fall after a frost (i.e., the 1-cut system) produces biomass low in livestock forage quality, but desirable quality for conversion to renewable fuels. Conversely, harvesting twice per year (i.e., the 2-cut system) increases switchgrass crop flexibility for landowners who may harvest the first cutting of switchgrass for hay or grazing when livestock forage quality is greatest [15] . Jacobs and King [15] noted that, averaged across all other treatment combinations, the 2-cut system produced significantly more biomass than the 1-cut system.

Despite the increased yield for switchgrass harvested in the 2-cut system, removing greater amounts of biomass also decreased near-surface extractable soil K and Mg levels. In addition, AS and mini-disk total infiltration rates were significantly lower in the 2-cut than in the 1-cut system. In a similar study, Edaño [34] noted switchgrass harvested in a 2-cut system had significantly lower extractable soil nitrate concentrations, total microbial biomass, and total mycorrhizal biomass in the top 10 cm than those in a 1-cut harvest system.

The lack of significant effects due to increased harvest frequency on root density are in contrast to harvest frequency effects on aboveground biomass yield. Jacobs and King [15] reported that mean biomass yield over the four-year study was significantly greater for the 2-cut (17.9 $Mg{\cdot}ha^{-1}$) compared to the 1-cut (12.1 $Mg{\cdot}ha^{-1}$) system when averaged over all other treatment factors. Data from above- and belowground plant production further demonstrate that increasing switchgrass yield by harvesting twice per year does not significantly decrease root density in the soil and climatic conditions of west-central Arkansas.

4.5. Infiltration Characteristics

In general, results indicated total infiltration rates were greater under tension when measured with the mini-disk infiltrometer compared to the double-ring infiltrometer because each method measured different mechanisms of infiltration. Infiltration was only measured in the inner ring for the double-ring infiltrometer while adjacent soil was nearly saturated throughout the 20-minute measurement period. Infiltration measured with the double-ring infiltrometer measured downward or vertical water infiltration (i.e., one-dimensional water infiltration). In contrast, soil around the mini-disk infiltrometer was unsaturated throughout the 20-minute measurement period; thus infiltration was likely both vertically downward and lateral (i.e., more three-dimensional water infiltration).

Due to the differences associated with measuring infiltration in unsaturated and nearly saturated conditions, only mini-disk infiltration rates were directly affected by harvest frequency. The mini-disk infiltration rate was greater in the 1-cut than in the 2-cut treatment, which supported the hypothesis that more intensive harvest frequencies decrease water infiltration through soil micropores. Increases in wheel traffic from tractors and other harvesting equipment in 2-cut systems may not consistently increase bulk density, as previous data showed, but may reduce soil micropore abundance and volume and consequently hydraulic conductivity. Removal of the switchgrass canopy cover during the growing season (i.e., at the June harvest in the 2-cut system) likely exposed more soil than in the 1-cut treatment and may have allowed soil micropores to become sealed or clogged by temporary crusting from precipitation or irrigation.

The slope of the linear regression relating the natural logarithm of the infiltration rate and the natural logarithm of time provided additional information about what could happen to surface water (i.e., precipitation or irrigation) in each treatment combination. Large (i.e., steep) slopes represented greater surface infiltration per unit time, while smaller (i.e., flatter) slopes represented slower surface infiltration per unit time. This relationship was readily apparent when comparing the slopes averaged over fertilizer source for double-ring infiltration, which differed by cultivar within harvest frequency (Figure 5). The treatment combinations with modest infiltration rates and a smaller slope characterizing the relationship between the natural logarithm of the infiltration rate and the natural logarithm of time included the "Cave-in-Rock"-2-cut and "Alamo"-1-cut combinations, which did not differ (Figure 5). Conversely, the slopes characterizing the relationship between the natural logarithm of the infiltration rate and the natural logarithm of time associated with the "Cave-in-Rock"-1-cut and "Alamo"-2-cut treatment combinations were larger and represented greater surface infiltration per unit time (Figure 5). A similar relationship was observed when the treatments were averaged over cultivar and differed by fertilizer source within harvest frequency, as the slope of the PL-1-cut treatment combination was smaller compared to the slope of the CF-1-cut treatment combination (Figure 5). One reason the slope was smaller for the PL treatment combination could be related to the presence of monovalent cations in the PL, particularly Na, which could have caused more surface soil dispersion than CF. The ecological implications for greater infiltration and storage of surface water in the soil profile over shorter periods are vast and may affect the amount of plant available water for crop growth, the amount of off-site nutrient loss, and the potential for erosion of topsoil.

The mean double-ring infiltration rate into a silt-loam soil, which was measured in this switchgrass production study after four years of continuous and consistent management, was less than that reported by Bonin et al. [35] for switchgrass harvested in a 1-cut system in Ohio (2.8 $mm \cdot min^{-1}$). Decreased infiltration rates recorded in this study may be due to a variety of site differences among the two studies, including different switchgrass plant population densities resulting from differing initial seeding rates (i.e., 11.2 and 4.4 kg pure live $seed \cdot ha^{-1}$ in Ohio and for this study, respectively), differences in soil surface texture (i.e., silty clay loam in Ohio and silt loam in this study), and/or differences in switchgrass stand age (i.e., seven years in Ohio and four years in this study) [35] . Another factor that may have contributed to decreased infiltration rates at this site compared to the Ohio site [35] could have been the hydrophobic properties of ash from burning as part of annual stand management activities. Switchgrass stubble was burned each March throughout the four-year treatment application period in this study, while the Ohio site was not burned at all during their study period [35] . Consequently, reducing the frequency of prescribed burns may increase infiltration rates.

Studies that evaluate soil aggregate stability may give ancillary information toward interpreting water infiltration rates because soil aggregate size influences soil pore space. In this study, though AS did not systematically differ between harvest frequencies, the total tension infiltration rate was greater in the 1-cut than in the 2-cut system, which supports the supposition of Edaño [34] that harvesting switchgrass in a 1-cut system has the potential to increase water infiltration rates. Results from this study indicate water infiltration rates do not consistently improve (i.e., increase) with greater AS. In addition, results show that treatment effects may be overlooked if only saturated water infiltration is evaluated.

4.6. Practical Applications

Producers and other private landowners engaged in feedstock production must make informed management decisions to maximize production while protecting soil and water resources. A first step in providing information to producers and landowners lies in pairing aboveground management strategies to maximize biomass yield, such as cultivar choice, irrigation management, harvest frequency, and choice of fertilizer, with belowground consequences.

Similar to the results of this study, Jacobs and King [15] also noted that each switchgrass cultivar responded differently to the 2-cut harvest frequency, even though mean annual production over the four-year period did not differ significantly between cultivars within harvest frequency averaged over fertilizer source. The "Cave-in- Rock"-2-cut treatment combination produced an average of 62% more biomass compared to the 1-cut system, while the "Alamo"-2-cut treatment combination only produced an average of 35% more biomass compared to the 1-cut system. The results regarding cultivar response to the 2-cut harvest frequency were similar to those reported by Fike et al. [36] , where maximum switchgrass biomass yield was achieved by harvesting lowland cultivars (e.g., "Alamo") in a 1-cut harvest frequency and upland cultivars (e.g., "Cave-in-Rock") in a 2-cut harvest frequency [36]. This study also supported the conclusion of Edaño [34] that lowland cultivars are better suited to dedicated energy crop production, while upland cultivars are better suited as dual-use livestock forage and energy crop production.

Biomass yield information coupled with soil hydraulic properties provide valuable information to landowners for making management decisions.

For example, landowners may choose to harvest twice per year only if "Cave-in-Rock" is used, knowing that increased biomass yield could be balanced by the trade-off of decreased water infiltration rates. Similarly, landowners may only harvest "Alamo" once per year, as harvesting twice per year may only increase yields by 35%, while also decreasing infiltration rates. Landowners who are interested in converting highly erodible cropland to switchgrass production may choose to plant "Alamo" instead of "Cave- in-Rock" to decrease the potential for soil erosion due to more water-stable soil aggregation. Subtleties associated with these management strategies may become important if switchgrass is grown on the vast areas needed to support industrial facilities for renewable fuel production.

5. CONCLUSIONS

Results showed that four years of consistent agronomic management with various strategies to maximize switchgrass production in west-central Arkansas produced significant soil property differences, some of which were small and of limited practical use, while other differences were larger and had practical management implications. Results from this study also illuminated new opportunities for producers and landowners to customize switchgrass management systems to address specific natural resource goals. In general, "Alamo" produced greater biomass yields and greater AS, which resulted in greater depletion of extractable soil P and K contents over time compared with "Cave-in-Rock". Irrigating switchgrass, though not currently cost-effective or recommended for marginal sites, significantly increased AS and switchgrass root densities when averaged over all other treatment factors. Harvesting switchgrass in a 2-cut system significantly decreased AS, extractable soil K and Mg contents, and total tension infiltration rates. Fertilizing switchgrass with PL rather than CF generally increased soil pH, EC, and extractable soil P, K, Ca and Mg contents, though both PL and CF appear to be viable options for landowners interested in switchgrass production.

Further investigations are needed to evaluate the effect of management strategies when no fertilizer is applied to assess soil property change if or when the bulk of fertilizers are diverted for growing food crops to feed a growing world population. More information is also needed about the fate of surface water and infiltration characteristics associated with switchgrass production strategies, as the importance of surface water and groundwater recharge only increase with time. Finally, this study demonstrated the need to enlarge the scope of discussion when assessing the feasibility and natural resource consequences of producing energy crops in the US.

ACKNOWLEDGEMENTS

The authors are grateful for the field and laboratory assistance provided by Bryan Jacobs, Eddie Pratt, Debbie Orick, Dale Goff, Michele Helton, Julie Osborne, and Taylor Adams.

REFERENCES

1. Energy Information Administration (EIA) (2013) How Dependent Are We on Foreign Oil?http://www.eia.gov/energy_in_ brief/ article/foreign_oil_dependence.cfm
2. United States Environmental Protection Agency (USEPA) (2013) Renewable Fuel Standard.http://www.epa.gov/otaq/ fuels/ renewablefuels/
3. Energy Information Administration (EIA) (2013) Monthly Energy Review.http://www.eia.gov/totalenergy/data/monthly/pdf/mer.pdf
4. Parrish, D.J. and Fike, J.H. (2005) The Biology and Agronomy of Switchgrass for Biofuels. Critical Reviews in Plant Science, 24, 423-459.

5. Tillman, D.A. (2000) Biomass Cofiring: The Technology, the Experience, the Combustion Consequences. Biomass & Bioenergy, 19, 365-384.

6. United States Department of Agriculture, Economic Research Service (USDA-ERS) (2012) Cattle and Beef Production Overview. http://www.ers.usda.gov/topics/animal-products/cattle-beef.aspx#.U-5by1bxVcw

7. Pimentel, D., Harvey, C., Resosidarmo, P., Sinclair, K., Kurz, D., McNair, M., Crist, S., Sphritz, L., Fitton, L., Saffouri, R. and Blair, R. (1995) Environmental and Economic Costs of Soil Erosion and Conservation Benefits. Science, 276, 1117-1123.

8. McLaughlin, S.B. and Kszos, L.A. (2005) Development of Switchgrass (Panicum virgatum) as a Bioenergy Feedstock in the United States. Biomass & Bioenergy, 28, 515-535.

9. Stubbendick, J., Hatch, S.L. and Butterfield, C.H. (1998) North American Range Plants. 4th Edition, University of Nebraska Press, Lincoln.

10. Popp, M.P. (2007) Assessment of Alternative Fuel Production from Switchgrass: An Example from Arkansas. American Journal of Agricultural Economics, 39, 373-380.

11. McLaughlin, S.B., Kiniry, J.R., Taliaferro, C.M. and De La Torre Ugarte, D. (2006) Projecting Yield and Utilization Potential of Switchgrass as an Energy Crop. Advances in Agronomy, 90, 267-297

12. United States Department of Agriculture, National Agricultural Statistics Service (USDA-NASS) (2014) Rank of States: Production of Livestock, Arkansas, 2013.http://www.nass.usda.gov/ Statistics_by_State/Arkansas/Publications/Statistical_Bulletin/Livestock_Data/arliverank.pdf

13. Mitchell, C.C. and Donald, J.O. (1999) The Value and Use of Poultry Manures as Fertilizer. Alabama Cooperative Extension System, Circular ANR-244, Alabama A&M University, Auburn University, Auburn.

14. Adler, P.R., Sanderson, M.A., Boateng, A.A., Weimer, P.J. and Jung, H.J.G. (2006) Biomass Yield and Biofuel Quality of Switchgrass Harvested in Fall and Spring. Agronomy Journal, 98, 1518-1525.

15. Jacobs, A.A. and King, J.R. (2012) Using Fertilization, Irrigation, and Harvest Strategies to Maximize "Alamo" and "Cave-in-Rock" Switchgrass Biomass Yield in the Southern Ozarks. http://www.plant-materials.nrcs.usda.gov/pubs/arpmctr11197.pdf

16. Kering, M.K., Butler, T.J., Biermacher, J.T., Mosali, J. and Guretzky, J.A. (2012) Effect of Potassium and Nitrogen Fertilizer on Switchgrass Productivity and Nutrient Removal Rates under Two Harvest Systems on a Low Potassium Soil. BioEnergy Research, 6, 329-335.

17. Bransby, D. and Huang, P. (2014) Twenty-Year Biomass Yields of Eight Switchgrass Cultivars in Alabama. BioEnergy Research, 7, 1186-1190.

18. Burner, D.M. (2012) Weather Summary for the Dale Bumpers Small Farms Research Center 2005-2012. Unpublished Data, Booneville.

19. National Oceanic and Atmospheric Administration (NOAA) (2010) Climatography of the United States, Annual and Seasonal Normals of Temperature and Precipitation 1981-2010: Booneville, Arkansas. United States Department of Commerce, National Climatic Data Center, Asheville.

20. United States Department of Agriculture, Natural Resource Conservation Service (USDA-NRCS) (2003) Leadvale Soil Series Official Description. https://soilseries.sc.egov.usda.gov/OSD_Docs/L/LEADVALE.html

21. United States Department of Agriculture, Natural Resource Conservation Service (USDA-NRCS) (2009) Establishing Native Warm Season Grass Mixtures Fact Sheet.http://efotg.sc.egov.usda.gov/references/public/AR/native_warm_season_grass.pdf

22. Bouyoucos, G.J. (1927) The Hydrometer as a New Method for the Mechanical Analysis of Soils. Soil Science, 23, 343- 353.

23. Mehlich, A. (1984) Mehlich 3 Soil Test Extractant: A Modification of Mehlich 2 Extractant. Communications in Soil Science & Plant Analysis, 15, 1409-1419.

24. Tucker, M.R. (1992) Determination of Phosphorous by Mehlich 3 Extraction. In: Donohue, S.J., Ed., Reference Soil and Media Diagnostic Procedure for the Southern Region of the United States, Southern Cooperative Series Bulletin 374, Virginia Agricultural Experiment Station, Blacksburg, 9-12.

25. Yoder, R.E. (1936) A Direct Method of Aggregate Analysis of Soils and a Study of the Physical Nature of Erosion Losses. Agronomy Journal, 28, 337-351.

26. Brye, K.R. and Riley, T.L. (2009) Soil and Plant Property Differences across a Chronosequence of Humid-Temperate Tallgrass Prairie Restorations. Soil Science, 174, 346-357.

27. Smith, S.F., Brye, K.R., Chen, P., Gbur, E. and Korth, K. (2014) Residue and Water Management Effects on Aggregate Stability and Aggregate-Associated Carbon and Nitrogen in a Wheat-Soybean, Double-Crop System. Soil Science Society of America Journal, 78, 1378-1391.

28. Anders, M.M., Brye, K.R., Olk, D.C. and Schmid, B.T. (2010) Rice Rotation and Tillage Effects on Soil Aggregation and Aggregate Carbon and Nitrogen Dynamics. Soil Science Society of America Journal, 76, 994-1004.

29. Frank, A.B., Berdahl, D.J., Hanson, J.D., Liebig, M.A. and Johnson, H.A. (2004) Biomass and Carbon Partitioning in Switchgrass. Crop Science, 44, 1391-1396.

30. Jung, J.Y., Lal, R., Jastrow, J.D. and Tyler, D.D. (2011) Nitrogenous Fertilizer Effects on Soil Structural Properties under Switchgrass. Agriculture, Ecosystems & Environment, 141, 215-220.

31. Six, J., Bossuyt, H., Degryze, D. and Denef, K. (2004) A History of Research on the Link between (Micro)aggregates, Soil Biota, and Soil Organic Matter Dynamics. Soil & Tillage Research, 79, 7-31.

32. Singer, M.J., Southard, R.J., Warrington, D.N. and Janitzky, P. (1992) Stability of Synthetic Sand-Clay Aggregates after Wetting and Drying Cycles. Soil Science Society of America Journal, 56, 1843-1848.

33. Whalen, J.K. and Chang, C. (2002) Macroaggregate Characteristics in Cultivated Soils after 25 Annual Manure Applications. Soil Science Society of America Journal, 66, 1637-1647.

34. Edaño, M.L.S. (2009) Effects of Nutrient Sources, Harvest Frequency and Environmental Conditions in Switchgrass Production and Soil Properties. Ph.D. Dissertation, Oklahoma State University, Stillwater.

35. Bonin, C., Lai, R., Schmitz, M. and Wullschleger, S. (2012) Soil Physical and Hydrological Properties under Three Biofuel Crops in Ohio. Acta Agriculturae Scandinavica, 62, 595-603.

36. Fike, J.H., Parrish, D.J., Wolf, D.D., Balasko, J.A., Green, J.T., Rasnake, M. and Reynolds, J.H. (2006) Switchgrass Production for the Upper Southeastern USA: Influence of Cultivar and Cutting Frequency on Biomass Yields. Biomass & Bioenergy, 30, 207-213.

CHAPTER 7

Response of Soil Microbial Properties to Long-Term Application of Organic and Inorganic Amendments in a Tropical Soil (Saria, Burkina Faso)

Ndeye Hélène Diallo-Diagne[1,2*], Komi Assigbetse[3,4], Saïdou Sall[5], Dominique Masse[3,4], Moussa Bonzi[6], Ibrahima Ndoye[7], Jean Luc Chotte[4]

[1]ISRA, LEMSAT, Centre IRD-ISRA Bel Air, Dakar, Sénégal

[2]Virginia Tech, Crop and Soil Environmental Sciences, Blacksburg, USA

[3]IRD, UMR Eco & Sols, LEMSAT, Centre IRD-ISRA Bel Air, Dakar, Sénégal

[4]IRD, UMR Eco & Sols, IRD, Montpellier, France

[5]UGB, UFR S2ATA, Saint Louis, Sénégal

[6]INERA, Ouagadougou, Burkina Faso

[7]UCAD, Dakar, Sénégal

ABSTRACT

Soil microbial biomass carbon (MBC), β-glucosidase, acid phosphatase and fluorescein diacetate (FDA) activities and bacterial community structure were assessed in a long-term (26 years) experiment, at physiological stages of sorghum growth, comparing different management methods for organic (manure, straw residues) and inorganic (urea) amendments at the INERA field station in Saria (Burkina Faso). Annual application of manure led to the highest soil microbial biomass and enzyme activities. Investigations indicated that only microbial biomass and β-glucosidase activities were affected during the cropping season. Phosphatase and FDA enzyme activities did not depend on the crop development stages. The application of N fertilizer modified phosphatase and FDA enzyme activities, the activities being higher in soils amended with N fertilizer. The bacterial community structure was analyzed by PCR-denaturing gradient gel electrophoresis (PCR-DGGE) targeting the eubacterial 16S rRNA gene. Cluster analysis of PCR-DGGE patterns showed two major clusters, the first containing the mineral fertilization and straw treatments and the second, the straw + urea, manure and manure + urea treatments. Sorghum grain yields were the highest for manure treatments. In this long-term experiment, applying straw did not produce a better grain yield than that obtained in the un-amended plot.

KEYWORDS

Organic Amendments, Mineral Fertilizer, Microbial Biomass, Soil Enzyme Activities, PCR-DGGE, Crop Development Stages, Lixisol

1. INTRODUCTION

In the Sudano-Sahelian region, agricultural soil productivity has been severely impaired owing to intensive cultivation, drought and soil erosion, which have led to a decrease in soil organic matter, a depletion of soil nutrients. Furthermore, in semi-arid agricultural conditions, crop residues are removed and organic input such as manure is rare. This leads to agricultural soil degradation with detrimental effects on the physical, chemical and biological properties of soil [1] .

To secure rural livelihoods, soil fertility and functioning must be restored to provide sustainable farming systems. As alternative, low cost solutions for increasing soil organic matter have been proposed. Bationo et al. [2]

and Badiane et al. [3] recommended regular inputs of organic amendments to soil while other authors advocated the use of improved fallows to restore organic matter to soil [4] [5] . These agricultural management methods are known to enhance the stock of soil carbon [6] . It is well established that organic matter plays a central role in maintaining key soil functions and is an essential determinant of soil fertility and resistance to erosion. Several authors have found that organic inputs increase soil microbial biomass [7] - [9] and enzyme activities [10] [11] and enhance microbial activity [12] . The application of organic amendments can, therefore, affect the activity, structure, diversity and function of soil microorganisms, which play a key role in soil processes to a greater extent than the application of inorganic fertilizers [7] [13] . All biochemical reactions in soils are mediated by the activities of enzymes and microorganisms are the major source of enzymes in soil. Microbial enzymes are involved in soil organic matter formation and decomposition and are used as indicators of soil fertility. Measurement of soil hydrolases provides an early indication of changes in soil fertility, since they are related to the mineralization of important nutrient elements and their dynamics in these systems may regulate nutrient uptake and plant growth. However, [14] suggested that some enzymes as b-glucosidase, and acid phosphatase are more sensitive to shifts in soil management and change more quickly than others. The enzyme b-Glucosidase is a common and predominant enzyme in soils and catalyzes the hydrolysis of cellobiose, and thus plays a major role in the initial phases of the decomposition of organic C compounds [15] . Acid phosphatase catalyzes the hydrolysis of a variety of organic phosphomonoesters and is therefore important in soil organic P mineralization and plant nutrition. FDA is hydrolysed by esterases, proteases and lipases, enzymes involved in the microbial decomposition of organic matter in soil [16] .

The recent development of genetic techniques targeting the 16S ribosomal DNA gene or functional genes has provided an alternative to culture-based methods, providing a unique insight into the composition, richness, structure and functioning of microbial communities [17] - [19] . Denaturing gradient gel electrophoresis (DGGE) of PCR-amplified genes is one of these genetic techniques and can be used to assess the diversity of complex microbial systems [20] [21] . The great interest in determining biodiversity and function of microorganisms in agrosystems and its utilization in soil system modeling is no longer to demonstrate but there is little information on the soil microbial communities as affected by management and land uses in semiarid conditions especially for tropical regions [22] - [24] where there is rapid turnover of soil organic matter.

This paper presents the impact of the long-term application of organic amendments and inorganic fertilizers on the activity of microbial communities and the genetic structure of the total bacterial community in a 26-year experiment in semi-arid conditions in Saria (Burkina Faso). The relationships between these properties, the amendments and yields are discussed.

2. MATERIALS AND METHODS

2.1. Study Site

The study was conducted in the Saria II site, a long-term field experimental site located at the Saria Agricultural Research Station (12°16'N, 2°9'W, 300 m altitude) in Burkina Faso. The climate is north-Sudanian [25] and rainfall is confined to the period from May to October with an annual mean of 800 mm. The main species of natural vegetation are Parkia biglobosa, Vitellaria paradoxa, Tamarindus indica, Andropogon gayanus and Pennisetum pedicellatum. The soil is Ferric Lixisol [26] . Soil organic carbon (SOC), N and available P in soil are very low (Table 1). Since there are no carbonates in the studied soil, total C was assumed to be organic C. The Cation Exchange Capacity (CEC) is poor and the soil is acid with a pH (H_2O) between 4.09 and 5.51.

2.2. Experimental Design

The experiment was set up in 1980. The treatments included in the present study are Con: control (plots without amendment); ConN: control + N fertilizer (urea, 60 kg · N · ha^{-1}); Str: sorghum straw residue (10 t · ha^{-1}); StrN: sorghum straw residue + urea (60 kg · N · ha^{-1}); Man: animal manure (10 t · ha^{-1}); ManN: animal manure + urea (60 kg · N · ha^{-1}). Each treatment was replicated six times in a randomized block design. Organic materials (manure and straw) were applied manually each year in July and incorporated into the soil. Each plot was 5.2 m × 4 m. A 2 m strip was left between each block. The plots were ploughed after the first significant rain event to a depth of 0.2 m. The cropping system is a continuous monoculture of sorghum (Sorghum bicolor L., cultivar Sariasso 14) with 31.250 plants ha^{-1} and 0.4 m × 0.8 m spacing between seed-beds. The sorghum grain yield was measured in all plots from an area of 9.6 m^2. The grain yield was obtained after air-drying until a constant weight was obtained.

2.3. Soil Sampling

Soil samples were taken after 26 years of continuous cultivation and at different stages of development of sorghum: on sowing (July), 6 weeks after sowing (August), during flowering (September) and on harvesting (November). The six replicate plots of each treatment were used as independent replicates for each analysis. For each plot, a composite soil sample was collected from the 0 - 10 cm soil layer. Field-moist soil samples were sieved (<2 mm) and further divided into two subsamples. The first was stored at 4°C for microbial biomass and enzyme activities measurements performed within one month following each sampling date. The other sub- samples were stored at −20°C for molecular analyses.

2.4. Microbial Biomass C

The microbial biomass C was determined from 20 g of fresh soil by the chloroform fumigation-extraction method [27] . The microbial biomass in fumigated (10 days) and unfumigated (control) soil was quantified using ninhydrin-N reactive compounds extracted from soils with 1 M KCl. The ninhydrin-reactive nitrogen content was determined using a continuous flow colorimeter (Evolution II, Alliance-Instrument, France).

The biomass C was calculated as the increase in ninhydrin-N between fumigated and unfumigated soils, multiplied by 21. The results were expressed as μg · C · g^{-1} soil.

2.5. Enzyme Assays

The β-glucosidase activity was assessed using 100 μl of 5 mM p-nitrophenyl-β-D-glucanopyranoside (PNG) as the substrate [28] . Soil subsamples (100 mg of fresh soil) were incubated at 37°C for 2 h with 400 μl of citrate phostphate buffer at pH 5.8. The p-nitrophenol (p-NP) produced in the enzymatic reactions was extracted and determined at 400 nm. The results were expressed as μg p-NP released g^{-1} · soil · h^{-1}. Acid phosphatase activity was determined using p-nitrophenyl phosphate disodium (p-NPP, 5 mM) as substrate [29] . Fresh soil (100 mg) was incubated at 37°C for 1 h with 0.5 M sodium acetate buffer at pH 6 and 100 μl of substrate [30] . The p-nitrophenol (p-NP) formed was determined by spectrophotometry at 400 nm and the results were expressed as μg p-NP released g^{-1} · soil · h^{-1}. The fluorescein diacetate activity (FDA) was determined using the method described by Adam and Duncan [31] . A 1 g-fresh soil was exposed to 200 μl of FDA solution (FDA 1 mg · $ml^{-1}$and phosphate potassium buffer, 60 mM) at pH 7.6 for 1 h at 30°C. Measurements were taken at 490 nm and the results were expressed as μg fluorescein released g^{-1} · soil · h^{-1}.

Table 1. Soil characteristics, plant biomass and grain yields of the differently amended plots at Saria II experimental station, Burkina Faso. Data are Means (Standard Error) of six replicates.

	Con	ConN	Str	StrN	Man	ManN
Total C (mg/g)	2.16 (0.3) b	2.22 (0.8) b	2.48 (0.0) b	2.68 (0.0) b	3.57 (0.0) a	3.56 (0.0) a
Total N (mg/g)	0.15 (0.0) b	0.16 (0.0) b	0.15 (0.0) b	0.17 (0.0) ab	0.27 (0.0) a	0.25 (0.0) ab
CEC (meq/%)	1.95 (0.3) a	2.24 (0.7) a	2.13 (0.6) a	1.99 (0.4) a	2.37 (0.4) a	2.33 (0.4) a
pH KCl	4.58 (0.3) c	4.09 (0.2) d	5.19 (0.2) ab	4.76 (0.2) c	5.51 (0.3) a	5.14 (0.4) b
Total P (mg/kg)	112.67 (17.3) b	96.83 (30.2) b	113.67 (15.1) b	112.67 (20.9) b	147.33 (19.2) a	146.50 (23.4) a
Available P (mg/kg)	30.77 (4.9) bc	28.22 (8.0) bc	31.64 (11.9) bc	26.19 (5.3) c	56.37 (28.1) a	45.53 (18.9) ab
Grain yield (kg/ha)	167 (174) b	207 (124) b	90 (60) b	233 (154) b	730 (319) a	737 (350) a
Plant biomass (kg/ha)	480 (426) b	627 (286) b	300 (192) b	600 (390) b	1893 (779) a	1900 (806) a

Means followed by the same letter in the same rows do not differ significantly ($P < 0.05$).

2.6. Bacterial Community Structure Analysis

Molecular analyzes were limited to 4 replicates per sample in order to reduce their related cost. Total DNA extraction was done as described by Hai et al. [32] . The extractions were performed using "Precellys-Keramik-Kit" lysing tubes (PEQLAB Biotechnologie GmbH, Germany) and the suspensions were then subjected to bead beating (BeadBeater, Biospec Products). DNA concentrations were measured and checked for the presence of potential PCR inhibitors using a NanoDrop fluorospectrophotometer (NanoDrop Technologies, Delaware, USA).

PCR-denaturing gradient gel electrophoresis (PCR-DGGE): The V3 region of the 16S rRNA gene was amplified for denaturing gradient gel electrophoresis analysis using primers UNIV518r [33] and EUB338f with a 40-bp GC clamp at the 5' end [34] . A touchdown PCR was performed in 25 μl containing 2.5 ng ADN, Taq Polymerase Ready-To-Go (Amersham-Bioscienes, France) and 0.5 μM of each primer. The thermal profile for amplification was as follows: 5 min at 94°C; 20 cycles of 30 s at 94°C, 30 s at 65°C (the annealing temperature decreased by 0.5°C at every cycle), and 1 min at 72°C. The additional 10 cycles were run at an annealing temperature of 55°C. The PCR products were electrophoresed on 1.5% (w/v) agarose gels and stained for 30 min with ethidium bromide (1 mg · L^{-1}). For DGGE analysis, the amplified products (20 μl) were separated with the Ingeny U-Phor system (Ingeny, Goes, The Netherlands) in 8% polyacrylamide (acrylamide-bisacrylamide [37.5:1]) gel containing a linear 45% to 70% denaturant gradient (100% denaturant contained 7 M urea and 40% formamide). Electrophoresis was carried out using 1 X Tris-acetate-EDTA buffer at 100 V and 60°C for 17 h. Gels were stained for 30 min with ethidium bromide (1 mg · L^{-1}) and destained in distilled water for 10 min. The banding patterns were then photographed using a Vilber Lourmat gel imaging system (Vilber Lourmat, France). Band detection and their intensity quantification were performed using TotalLab gel imaging software (Nonlinear Dynamics Ltd., Newcastle UK), with manual checking and adjustment of each band position. The richness index (R) was calculated with the assumption that each band was an operational taxonomic unit (OTU). Band intensity was used as an indication of relative abundance. The clustering of the DGGE patterns analyzed using

TotalLab was performed using the Unweighted Pair Group method with Mathematical Average (UPGMA). The structural diversity of the microbial community was estimated by the Shannon index of general diversity [35] based on the number of bands present and the relative intensities of the bands in each lane. The intensity of the bands was estimated from the peak heights. The Shannon H index was calculated using the following equation:

$$H = -\Sigma(Pi * \log Pi)$$

where Pi is ni/N; ni is the peak height and N is the sum of all peak heights.

2.7. Statistical Analyses

The differences between treatments were estimated by three-way analysis of variance using XLSTAT 2010 software and the Fischer Least Significant Difference (LSD) test at 5% level used to determine whether the sampling date, treatment, nitrogen input and possibly interaction between them had any significant influence. Pearson correlation coefficients were used to assess relationships between selected soil variables. All results were significant when $P < 0.05$.

3. RESULTS

3.1. Sorghum Grain Yields and Plant Biomass

Grain yields ranged from 167 $kg \cdot ha^{-1}$ to 737 $kg \cdot ha^{-1}$ (Table 1). Significantly higher yields were obtained in soils amended with manure and manure + N-fertilizer ($P < 0.001$). No significant difference was noticed between straw-amended soils (Str and StrN) and control soils

(Con and ConN). Highest plant biomass yields (Table 1) were obtained from the manure-amended plots ($P < 0.001$). Interestingly, straw amendment reduced plant biomass yields (300 kg · ha^{-1}) compared to the untreated control (480 kg · ha^{-1}).

3.2. Microbial Biomass and Enzyme Activities

The results regarding microbial biomass and enzyme activities are shown in Table 2. The three-way analysis of variance carried out on the data allowed us to show the effect of the organic management, N fertilizer and the sampling date on microbial biomass and enzyme activities. There was an interaction between the sampling date and the organic management on microbial biomass and β-glucosidase activity. Application of N fertilizer had not modified these parameters. Overall, organic management had an effect at the first two sampling dates, at the time of sowing and 6 weeks later. In these cases values of microbial biomass were significantly ($P < 0.05$) higher than that of the controls and the assessment of the differences show: 44% for straw and 79% for manure at the time of sowing and 40% for straw and 62% for manure 6 weeks after sowing. The microbial biomass dropped considerably from flowering to an average of 6.01 μg · g^{-1} at harvest, a factor of 6.5 times less than the overall average measured at the time of sowing. β-glucosidase activity increased significantly ($P < 0.05$) with the addition of manure or straw with respect to the control at the first three sampling dates. This difference was no longer significant at the final sampling (harvest).

Regarding phosphatase and FDA activities, there was no significant interaction between the date, organic management and N fertilizer application. The maximum values for these variables were for samples taken at the time of sowing and, above all, during flowering. Adding manure also resulted in an average increase of 49% for phosphatase and 104% for FDA relative to the control, regardless of the sampling date or the addition of N fertilizer. Adding straw had a lesser effect with an average increase of 33% for phosphatase and 59% for FDA relative to the control. Adding N increased phosphatase by 13% and FDA by 22% for FDA relative to plots without N fertilizer addition.

Table 2. The effect of long term fertilization on microbial biomass and enzymatic activities in soils collection at the four sampling seasons. Data are means (standard error) of six replicates

Date	Treatment					
	Con	ConN	Str	StrN	Man	ManN
Microbial biomass (mg·C·kg^{-1}·soil)						
	Con	ConN	Str	StrN	Man	ManN
Sowing	29.72 (0.87)	25.88 (1.81)	41.6 (2.84)	38.44 (1.42)	49.88 (1.07)	49.88 (1.41)
6 weeks	33.05 (1.77)	31.00 (1.27)	47.11 (2.43)	42.38 (2.78)	52.55 (1.44)	51.51 (1.33)
Flowering	17.27 (2.72)	26.16 (2.10)	11.61 (1.18)	27.55 (3.62)	15.33 (3.69)	19.77 (2.08)
Harvested	4.94 (0.44)	5.88 (0.67)	6.16 (0.65)	4.16 (0.30)	6.5 (0.70)	8.6 (0.87)
Phosphatase (μg·PNP·g^{-1}·soil·h^{-1})						
Sowing	85.87 (2.0)	92.23 (4.5)	119.29 (4.0)	127.72 (4.2)	127.75 (5.0)	146.03 (4.0)
6 weeks	85.34 (4.9)	85.73 (2.5)	109.34 (2.8)	112.42 (4.5)	109.35 (1.3)	141.89 (5.4)
Flowering	91.02 (4.6)	102.86 (7.7)	113.33 (5.4)	146.80 (7.6)	124.53 (7.4)	158.89 (7.1)
Harvested	73.20 (2.8)	91.96 (4.3)	104.17 (3.4)	109.75 (2.8)	114.16 (4.3)	131.75 (3.4)
B-glucosidase (μg·PNP·g^{-1}·soil·h^{-1})						
Sowing	9.87 (0.7)	6.70 (0.6)	14.55 (0.5)	12.49 (0.3)	15.71 (0.4)	15.62 (0.5)
6 weeks	8.41 (0.6)	7.24 (1.0)	14.09 (0.6)	14.34 (0.6)	18.05 (0.7)	16.85 (0.7)
Flowering	9.82 (0.8)	6.21 (0.5)	14.46 (0.8)	19.76 (0.8)	22.13 (1.3)	18.97 (0.6)
Harvested	16.84 (0.9)	21.27 (2.0)	17.29 (0.7)	16.95 (1.3)	24.61 (0.7)	19.30 (0.7)
Fluorescein diacetate (μg·fl·g^{-1}·soil·h^{-1})						
Sowing	0.86 (0.0)	0.81 (0.0)	1.32 (0.0)	1.80 (0.1)	1.37 (0.0)	2.07 (0.0)
6 weeks	1.35 (0.1)	1.85 (0.1)	2.65 (0.1)	3.25 (0.1)	3.65 (0.1)	3.79 (0.1)
Flowering	1.30 (0.0)	1.75 (0.0)	1.63 (0.1)	2.70 (0.1)	2.94 (0.2)	3.20 (0.2)
Harvested	1.17 (0.0)	1.74 (0.1)	1.84 (0.1)	1.98 (0.1)	2.51 (0.1)	2.59 (0.0)

Source of variation	*df*	Microbial Biomass	Phosphatase	B-glucosidase	Fluorescein diacetate
			P (<0.05)		
Date	3	<0.0001	0.037	<0.0001	<0.0001
Input management	2	0.000	<0.0001	<0.0001	<0.0001
N fertilizer	1	ns	0.002	ns	0.002
Date*input management	6	0.001	ns	0.024	ns
Date*N fertilizer	3	ns	ns	ns	ns
Input management*N fertilizer	2	ns	ns	ns	ns
Date*input management*N fertilizer	6	ns	ns	ns	ns

3.3. Bacterial Community Structure

The DGGE of the 16S rRNA amplified from all of the treatments displayed different patterns. The clustering method revealed two major clusters at sowing, 6 weeks and harvest sampling time (Figure 1). Cluster 1 groups the soil amended with manure (Man and ManN) and straw plus

N fertilizer (StrN), while the control (Con, and ConN) and soil amended with straw (Str) are grouped in cluster 2. However, this differentiation was less evident for samples taken during flowering. There was less difference in the community structure in StrN and Man amended soils than in the straw and control treatments. The patterns had some common and dominant bands indicating that there were common bacterial types in the soils regardless of the type of amendment applied and the plant development stage.

Globally the richness R index and the Shannon-Weaver diversity index (H) had the same trend and showed that the microbial diversity varied with the type of soil management and generally increased with the application of manure (Table 3). Microbial diversity varied significantly ($P < 0.05$) with the sorghum development stages with the lowest values recorded at flowering.

3.4. Correlation between the Microbial Biomass, Enzyme Activities and Shannon's Diversity

Correlation coefficients between microbial biomass, enzyme activities and the Shannon diversity were calculated (Table 4) and revealed distinct variation during the different sorghum development stages. At sowing a negative correlation was observed between Shannon's diversity and microbial biomass ($P = 0.003$) and FDA ($P = 0.011$) whereas positive correlations were found between microbial biomass and phosphatase ($P < 0.0001$), FDA ($P < 0.0001$), β-glucosidase ($P = 0.000$) activities at this date and 6 weeks sampling date. At flowering the Shannon's diversity and enzyme activities (phosphatase $P = 0.002$), FDA ($P = 0.001$), β-glucosidase ($P = 0.007$) were correlated. At harvest time only FDA and Phosphatase were positively correlated ($P < 0.0001$).

4. DISCUSSION

Long-term fertilization had a significant effect on soil organic C, total N content, total P and pH. The highest values of soil organic C, total N content, total P were recorded in soils amended with organic manure. Similar effects were shown in previous long-term field studies using comparable organic amendments. Masto et al. [10] , Chu et al. [7] and Liu et al. [23] showed that organic manure significantly increased the amount of soil organic C and total N. Mandal et al. [36] and Enwall et al. [37] on a long-term application site (34 and 51 years) have recorded a significant increase in organic C and total N for amendment with farmyard manure with NPK. The application of organic amendments (manure or sewage sludge) generally increased the SOM content to a much greater extent than that of inorganic fertilizer [23] [37] [38].

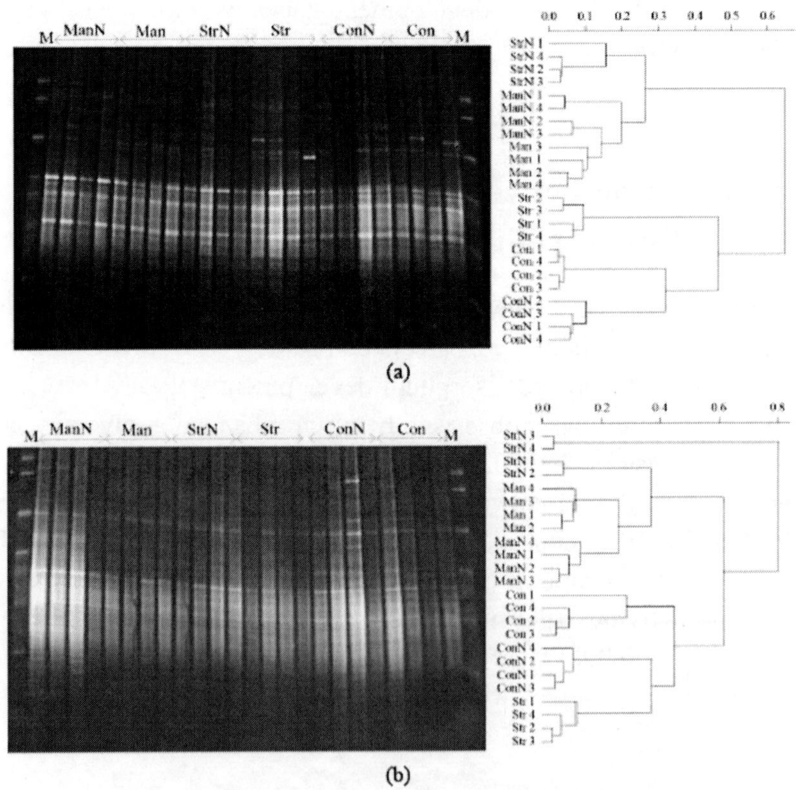

Figure 1. Profiles and UPGMA dendrogram describing pattern similarities of DGGE profiles obtained from treatments at different growth stages. (a) sowing, (b) 6 weeks after sowing, (c) flowering, (d) harvested; Con: control, ConN: control + urea, Str: straw, StrN: straw + urea, Man: manure, ManN: manure + urea.

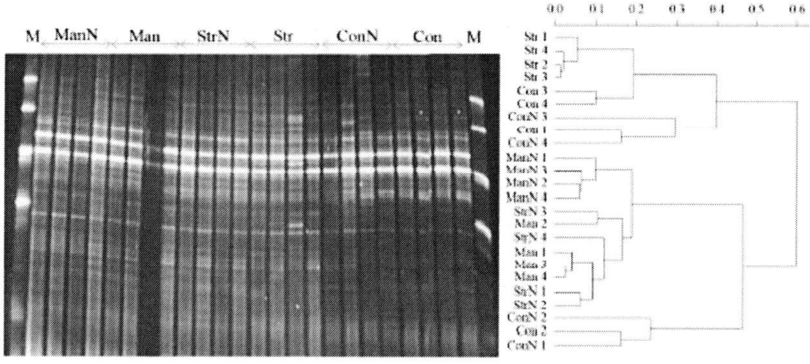

(c)

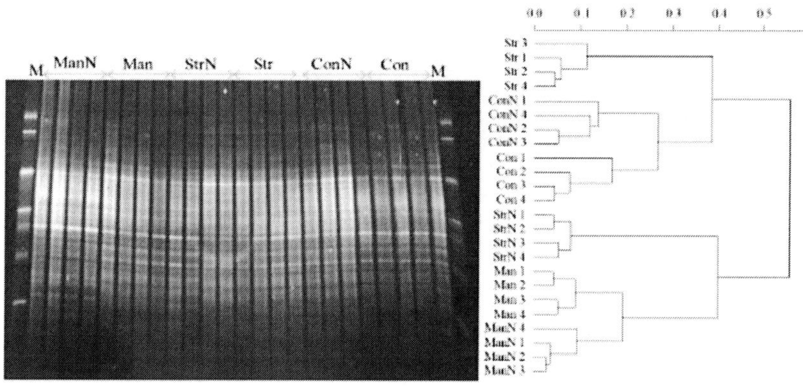

(d)

Figure 1. (Continued)

Table 3. Species richness (R) and Shannon diversity (H) indexes values for DGGE profiles from soils collection at the four sampling date.

	R				H			
	Sowing	6 weeks	Flowering	Harvested	Sowing	6 weeks	Flowering	Harvested
Con	^{B}23.00 c	^{A}25.75 b	^{C}13.50 d	^{B}22.50 c	^{A}1.30 c	^{A}1.31 bc	^{B}1.08 c	^{A}1.32 b
ConN	^{A}25.00 a	^{A}24.00 c	^{C}13.25 d	^{B}20.25 d	^{A}1.36 a	^{A}1.35 ab	^{C}1.06 c	^{B}1.27 c
Str	^{A}25.00 a	^{B}22.00 d	^{D}15.00 c	^{C}20.50 d	^{B}1.34 ab	^{C}1.29 cd	^{D}1.14 b	^{A}1.40 a
StrN	^{B}23.75 b	^{C}20.50 e	^{C}21.00 b	^{A}27.00 a	^{A}1.33 ab	^{C}1.25 d	^{C}1.24 a	^{B}1.28 c
Man	^{B}23.00 c	^{A}28.00 a	^{C}19.75 b	^{B}24.00 b	^{C}1.31 bc	^{A}1.36 a	^{D}1.23 a	^{B}1.33 b
ManN	^{C}22.50 d	^{A}28.75 a	^{C}22.50 a	^{B}24.50 b	^{C}1.28 c	^{A}1.39 a	^{C}1.27 a	^{B}1.34 b

Means ($n = 4$) within columns followed by the same letter are not significantly different ($P < 0.05$). Capital letters indicate significant difference in the same row ($P < 0.05$).

Table 4. Correlation matrix of pearson (P < 0.05) of microbial biomass, enzymes activities and Shannon index diversity enzymes activities.

Variables	Shannon H	Biomasse	Phosphatase	B-glucosidase	FDA
Sowing					
Shannon H	1				
Biomasse	-0.65^{***}	1			
Phosphatase	ns	**0.75^{***}**	1		
B-glucosidase	ns	0.67^{***}	0.61^{**}	1	
FDA	-0.49^{*}	**0.78^{***}**	**0.85^{***}**	0.49^{*}	1
6 weeks					
Shannon H	1				
Biomasse	ns	1			
Phosphatase	ns	0.71^{***}	1		
B-glucosidase	ns	0.46^{*}	0.62^{**}	1	
FDA	ns	0.74^{***}	0.73^{***}	0.71^{***}	1
Flowering					
Shannon H	1				
Biomasse	ns	1			
Phosphatase	0.56^{**}	ns	1		
B-glucosidase	0.65^{**}	ns	0.57^{**}	1	
FDA	0.50^{*}	ns	ns	ns	1
Harvest					
Shannon H	1				
Biomasse	ns	1			
Phosphatase	ns	ns	1		
B-glucosidase	ns	ns	ns	1	
FDA	ns	ns	**0.86^{***}**	ns	1

Coefficients higher than 0.70 are in bold NS Not significant, $^{*}P < 0.05$, $^{**}P < 0.01$, $^{***}P < 0.001$.

This study showed that N input through amendment with urea significantly decreased the soil pH. This could be attributable to the repeated application of inorganic fertilizers as is the case in Saria. This acidifying effect of N-fertilizer is in agreement with the findings of Hati et al. [39] and Liu et al. [23] . As agricultural use of inorganic fertilizers unavoidably increases soil acidity, Ge et al. [40] recommended that the combined use of organic manure with inorganic fertilizers should be considered depending on the balance between crop demand and the supply of available nutrients in the soil.

Microbial biomass has often been used to assess the effect of different farming systems on soil fertility [41] . In this study, the microbial biomass (MBC) varied according to the sorghum developmental stage.

High values were recorded at sowing and 6 weeks after sowing, which were periods with heavy rainfall. Long-term application of organic manure significantly increased the biomass C at the various sampling times, followed by amend- ment with straw. This result is consistent with previous findings of Gu et al. [9] and Liu et al. [23] who observed that the microbial biomass was considerably greater in soils receiving manure along with NPK fertilizer than in plots receiving only inorganic fertilizer. This increase in microorganisms might be explained by the source of easily degradable compounds through the addition of organic matter in the oligotrophic soil environment [42] . Previous studies [38] [43] reported that farming systems using green manures, manure and crop rotation to provide carbon inputs produced higher total microbial biomass than the systems using only inorganic fertilizers. These carbon rich materials can sustain microbial activity and growth, thus enhancing biogeochemical nutrient cycles [44] . The decreased MBC content after inorganic fertilizer application is also consistent with several previous reports [22] [45] [46] ; it could be explained by the acidifying effect of mineral N.

Houot and Chaussod [47] found a positive correlation between microbial biomass and C content on a 80-year corn/beet rotation site with various types and levels of fertilization. Organic amendment was also reported to increase enzyme activities in soil [48] [49] . The presence and activity of living beings in the soil result in the synthesis of enzymes. Enzymes play a key role in the nutrient cycle. Our results showed that extracellular enzymes (acid phosphatase, β-glucosidase, FDA) were enhanced by organic amendments, especially manure. The differences in these activities may be related to the soil properties such as carbon content and microbial biomass [10] [24] . The enzyme activities studied in this work (phosphatase, β-glucosidase and fluorescien diacetate) had different activity peaks: 6 weeks after sowing for FDA, flowering for phosphatase and on harvesting for β-gluco- sidase. This could be explained by the nature of their substrates. The FDA hydrolysis rate is widely accepted as an accurate method for measuring total microbial activity in soils [31] because FDA hydrolysis is mediated simultaneously by protease, esterase and lipase and it can reflect the activities of these enzymes in soil [50] [51] . FDA activity showed clear seasonal variation with the highest value 6 weeks after sowing correlated with the highest microbial biomass. The reduction of FDA on flowering and harvest may be attributed to the lower uptake of fluorescein diacetate by soil microbial cells owing to high drying stress and the destruction of enzymes such as protease. The β-glucosidase and phosphatase activities in soils have been found to be persistent for long periods owing to some form of enzyme protective mechanism existing in soils, for example bound to soil colloids and humic substances, free in soil solutions or associated with living or dead plants or microbial cells. These mechanisms may protect those enzymes against external factors such as desiccation. Our result suggests enrichment in soil litter of a cellulolitic [52] which acts as substrate for the enzymes at a later stage.

Molecular fingerprinting (DGGE) was used to examine the effects of long-term fertilization regimes on bacterial community structure. These results showed that long-term organic and inorganic fertilization had different effects on the genetic structure of the total bacterial community. The different treatments displayed different DGGE profiles. Nannipieri et al. [53] defined microbial diversity as a general term used to include genetic diversity, that is, the amount and distribution of genetic information, within microbial species. This DGGE technique allows to analyze many samples at a same time and to compare the communities during environmental modifications [20] [54] . DGGE technique take account alive and dead microorganisms that can cause extracellular enzymes found in the soil of which a correlation Shannon and enzyme activities is possible. The correlation between soils parameters showed that soil functioning and diversity vary according to the plant growth steps. Microbial biomass and enzyme activities were correlated at sowing and at the early stage of plant development. Greater enzyme activities could be related to greater production of enzymes by the increased microbial biomass [55] . Positive correlations between them support this view. At flowering, the correlations between diversity and microbial activities suggest that for this period known for much exudation and rhizodeposition, microbial activity could be linked to the diversity of microbial communities and not the density. Results showed that no direct relationship exists between the microbial properties, activities and community structure. However, the differences in response to the amendments suggest that microbial function and community structure reflect the microbial diversity. The differences in the microbial activity or function might be reflected in changes in community structure.

Changes in microbial activity and composition can in turn influence plant growth by increasing nutrient turnover. In practice, the crop yield is the result of all the physical, chemical, biological interactions in the soil. Results showed that the manure-amended plots displayed the highest grain yield indicating the benefits of organic manure on crop growth. These results are consistent with those obtained in long-term fertilizer experiments by Hati et al. [39] , Bi et al. [56] and Liu et al. [23] . However, straw-amendment did not stimulate grain yield. Moreover, no significant difference was measured between N fertilized and unfertilized treatments. Mando et al. [24] , investigating the same field experiment, obtained different results when comparing grain yields measured for the various treatments in 1999, 2000, 2001 and 2002. Their results also showed that soil amended with manure exhibited the highest grain yields, but they found that the application of straw led to higher grain biomass than that of the control soil. Moreover, they observed a significant positive impact of the application of N fertilizer in combination with straw. It seemed that the positive impact of straw recorded up to 2002 was no longer significant compared to un-amended soil in 2006. This result may be explained by the fact that organic materials such as cereal residues may have a strong influence on the immobilization of inorganic N [57] [58] and may release allelopathic compounds in soils [59] [60] , thereby affecting crop yields. A reduction in yield following straw addition has been reported in several works [61] [62].

5. CONCLUSION

This study, in one of the few long-term experimental sites in a tropical environment with semi-arid climate, complements previously published works [24] . It provides additional information by comparing the effect of repeated annual amendments of crop residues and manure with or without urea on the soil biological components. The higher microbial abundance, enzyme activities and soil bacteria diversity recorded with manure application reflected a better substrate and nutrient cycling capacity compared to other treatments. Long term fertilization of sorghum straw does not reveal an improvement in grain yields compared to mineral fertilized and control plots. This result shows that it is important to characterize the processes for release and mineralization of nutrients more clearly depending on the nature of the resources available for farmers.

ACKNOWLEDGEMENTS

This research was supported by ANR funds (ANR-Biodiversité-2005, Programme Microbes). We gratefully acknowledge Dr Armand koné for constructive comments on the manuscript.

REFERENCES

1. Rivero, C., Chirenje, T., Ma, L.Q. and Martinez, G. (2004) Influence of Compost on Soil Organic Matter Quality under Tropical Conditions. Geoderma, 123, 355-361.
2. Bationo, A., Lompo, F. and Koala, S. (1998) Research on Nutrient Flows and Balances in West Africa: State-of-the-Art. In: Smaling, E.M.A., Ed., Nutrient Balances as Indicators of Production and Sustainability in Sub-Saharan African Agricultural, Agriculture, Ecosystems & Environment, 71, 19-36.
3. Badiane, A.N., Khouma, M. and Sene, M. (2000) Région de Diourbel: Gestion des sols. Drylands Research Working Paper 15, Drylands Research, Somerset, 25 p.
4. Floret, C. and Pontanier, R. (1993) Recherches sur la jachère en Afrique tropicale, In: Floret, C., Pontanier, R. and Serpentié, G., Eds., La Jachère en Afrique Tropicale, Mab UNESCO, 11-54.
5. Manlay, R.J. (2000) Organic Matter Dynamics in Mixed-Farming Systems of the West African Savanna: A Village Case Study from South Senegal. Ph.D. Thesis, ENGREF, Paris, 278 p.
6. Roldan, A., Salinas-Garcia, J.R., Alguacil, M.M. and Caravaca, F. (2005) Changes in Soil Enzyme Activity, Fertility, Aggregation and

C Sequestration Mediated by Conservation Tillage Practices and Water Regime in a Maize Field. Applied Soil Ecology, 30, 11-20.

7. Chu, H.Y., Lin, X.G., Fujii, T., Morimotob, S., Yagi, K., Hu, J.L. and Zhang, J.B. (2007) Soil Microbial Biomass, Dehydrogenase Activity, Bacterial Community Structure in Response to Long-Term Fertilizer Management. Soil Biology & Biochemistry, 39, 2971-2976.

8. Koné, W.A., Tondoh, E.J., Angui, K.T.P., Bernhard-Reversat, F., Loranger-Merciris, G., Brunet, D. and Brédoumi, K.S.T. (2008) Is Soil Quality Improvement by Legume Cover Crops a Function of Initial Soil Chemical Characteristics? Nutrient Cycling in Agroecosystems, 82, 89-105.

9. Gu, Y., Zhang, X., Tu, S. and Lindstrom, K. (2009) Soil Microbial Biomass, Crop Yields, and Bacterial Community Structure as Affected by Long-Term Fertilizer Treatments under Wheat-Rice Cropping. European Journal of Soil Biology, 45, 239-246.

10. Masto, R.E., Chhonkar, P.K., Singh, D. and Patra, A.K. (2006) Changes in Soil Biological and Biochemical Characteristics in a Long-Term Field Trial on a Sub-Tropical Inceptisol. Soil Biology & Biochemistry, 38, 1577-1582.

11. Koné, W.A., Edoukou, E.F., Gonnety, T.J., N'Dri, A.A.N., Assémien, E.F.L., Angui, K.T.P. and Tondoh, E.J. (2012) Can the Shrub Chromolaena odorata (Asteraceae) Be Considered as Improving Soil Biology and Plant Nutrient Availability. Agroforestry Systems, 85, 233-245.

12. Hungria, M., Franchini, J.C., Brandao-Junior, O., Kaschuk, G. and Souza, R.A. (2009) Soil Microbial Activity and Crop Sustainability in a Long-Term Experiment with Three Soil-Tillage and Two Crop-Rotation Systems. Applied Soil Ecology, 42, 288-296.

13. Lejon, D.P.H., Sebastia, J., Lamy, I., Chaussod, R. and Ranjard, L. (2007) Relationships between Soil Organic Status and Microbial Community Density and Genetic Structure in Two Agricultural Soils Submitted to Various Types of Organic Management. Microbial Ecology, 53, 650-663.

14. Green, V.S., Stott, D.E., Cruz, J.C. and Curi, N. (2007) Tillage Impact on Soil Biological Activity and Aggregation in a Brazilian Cerrado Oxisol. Soil & Tillage Research, 92, 114-121.

15. Acosta-Martinez, V., Cruz, L., Sotomayor-Ramirez, D. and Perez-Alegria, L. (2007) Enzyme Activities as Affected by Soil Properties and Land Use in a Tropical Watershed. Applied Soil Ecology, 35, 35-45.

16. Schnürer, J. and Rosswall, T. (1982) Fluorescein Diacetate Hydrolysis as a Measure of Total Microbial Activity in Soil and Litter. Applied and Environmental Microbiology, 43, 1256-1261.

17. Van Elsas, J.D., Duarte, G.F., Rosado, A.S. and Smalla, K. (1998) Microbiological and Molecular Biological Methods for Monitoring Microbial Inoculants and Their Effects in the Soil Environment. Journal of Microbiological Methods, 32, 133-154.

18. Nüsslein, K. and Tiedje, J.M. (1998) Characterization of the Dominant and Rare Members of a Young Hawaiian Soil Bacterial Community with Small-Subunit Ribosomal DNA Amplified from DNA Fractionated on the Basis of Its Guanine and Cytosine Composition. Applied and Environmental Microbiology, 64, 1283-1289.

19. Griffiths, R.I., Whiteley, A.S., O'Donnell, A.G. and Bailey, M.J. (2003) Rapid Method for Coextraction of DNA and RNA from Natural Environments for Analysis of Ribosomal DNA and rRNA-Based Microbial Community Composition. Applied and Environmental Microbiology, 66, 5488-5491.

20. Ranjard, L., Poly, F. and Nazaret, S. (2000) Monitoring Complex Bacterial Communities Using Culture-Independent Molecular Techniques: Application to Soil Environment. Research in Microbiology, 151, 167-177.

21. Stark, C.H., Condron, L.M., O'Callaghan, M., Stewart, A. and Di, H.J. (2008) Differences in Soil Enzyme Activities, Microbial Community Structure and Short-Term Nitrogen Mineralisation Resulting from Farm Management History and Organic Matter Amendments. Soil Biology & Biochemistry, 40, 1352-1363.

22. Kaur, K., Kapoor, K.K. and Gupta, A.P. (2005) Impact of Organic Manures with and without Mineral Fertilizers on Soil Chemical and Biological Properties under Tropical Conditions. Journal of Plant Nutrition and Soil Science, 168, 117-122.

23. Liu, E., Yan, C., Mei, X., He, W., Bing, S.H., Ding, L., Liu, Q., Liu, S. and Fan, T. (2010) Long-Term Effect of Chemical Fertilizer, Straw, and Manure on Soil Chemical and Biological Properties in Northwest China. Geoderma, 158, 173-180.

24. Mando, A., Ouattara, B., Sédogo, M., Stroosnijder, L., Ouattara, K., Brussaard, L. and Vanlauwe, B. (2005) Long-Term Effect of Tillage and Manure Application on Soil Organic Fractions and Crop Performance under Sudano-Sahelian Conditions. Soil & Tillage Research, 80, 95-101.

25. Fontes, J. and Guinko, S. (1995) Carte de la végétation et de l'occupation du sol du Burkina Faso. Note Explicative. Ministère de la coopération francaise, Toulouse.

26. FAO (2006) World Reference Base for Soil Resources 2006: A Framework for International Classification, Correlation and Communication, World Soil Resources Reports 103. FAO, Rome.

27. Amato, M. and Ladd, J.N. (1988) Assay for Microbial Biomass Based on Ninhydrin-Reactive Nitrogen in Extracts of Fumigated Soils. Soil Biology & Biochemistry, 20, 107-114.

28. Hayano, K. (1973) A Method for Determination of β-Glucosidase Activity in Soil. Soil Science and Plant Nutrition, 19, 103-108.

29. Tabatabai, M.A. and Bremmer, J.M. (1969) Use of P-Nitrophenylphosphate for Essay of Soil Phosphatase Activity. Soil Biology & Biochemistry, 1, 301-307.

30. Nannipieri, P., Ceccanti, C., Ceverlli, S. and Matarese, E. (1980) Extraction of Phosphatase, Urease, Protease, Organic Carbon and Nitrogen from Soil. Soil Science Society of America Journal, 44, 1011-1016.

31. Adam, G. and Duncan, H. (2001) Development of a Sensitive and Rapid Method for the Measurement of Total Microbial Activity

Using Fluorescein Diacetate (FDA) in a Range of Soils. Soil Biology & Biochemistry, 33, 943-951.

32. Hai, B., Diallo, N.H., Sall, S., Felix, H., Schauss, K., Bonzi, M., Assigbetse, K., Chotte, J.L., Munch, J.C. and Schloter, M. (2009) Quantification of Key Genes Steering the Microbial Nitrogen Cycle in the Rhizosphere of Sorghum Cultivars in Tropical Agroecosystems. Applied and Environmental Microbiology, 75, 4993-5000.

33. Ovreas, L., Forney, L., Daae, F.L. and Torsvik, V. (1997) Distribution of Bacterioplankton in Meromitic Lake Saelenvannet, as Determined by Denaturing Gradient Gel Electrophoresis of PCR-Amplified Gene Fragments Coding for 16S rRNA. Applied and Environmental Microbiology, 63, 3367-3373.

34. Muyzer, G., Waal, E.C.D. and Uitterlinden, A.G. (1993) Profiling of Complex Microbial Populations by Denaturing Gradient Gel Electrophoresis Analysis of Polymerase Chain Reaction-Amplified Genes Coding for 16S rRNA. Applied and Environmental Microbiology, 59, 695-700.

35. Shannon, C.E. and Weaver, W. (1963) The Mathematical Theory of Communication. University of Illinois Press, Urbana.

36. Mandal, A., Patra, A.K., Singh, D., Swarup, A. and Masto, R.E. (2007) Effect of Long-Term Application of Manure and Fertilizer on Biological and Biochemical Activities in Soil during Crop Development Stages. Bioresource Technology, 98, 3585-3592.

37. Enwall, K., Nyberg, K., Bertilsson, S., Cederlund, H., Stenstrom, J. and Hallin, S. (2007) Long-Term Impact of Fertilization on Activity and Composition of Bacterial Communities and Metabolic Guilds in Agricultural Soil. Soil Biology & Biochemistry, 39, 106-115.

38. Wen, Y.L., Xiao, J., Li, H., Shen, Q.R., Ran, W., Zhou, Q.S. and Yu, G.H. (2014) Long-Term Fertilization Practices Alter Aluminnjum Fractions and Coordinate State in Soil Colloids. Soil Science Society of America Journal, 78, 2083-2089.

39. Hati, K.M., Swarup, A., Mishra, B., Manna, M.C., Wanjari, R.H., Mandal, K.G. and Misra, A.K. (2008) Impact of Long-Term Application of Fertilizer, Manure and Lime under Intensive Cropping on Physical Properties and Organic Carbon Content of an Alfisol. Geoderma, 148, 173-179.

40. Ge, G., Li, Z., Fan, F., Chu, G., Hou, Z. and Liang, Y. (2010) Soil Biological Activity and Their Seasonal Variations in Responsible to Long-Term Application of Organic and Inorganic Fertilizers. Plant and Soil, 326, 31-44.

41. Hassink, J., Lebbink, G. and Van Veen, J.A. (1991) Microbial Biomass and Activity of a Reclaimed-Polder Soil under a Conventional or a Reduced-Input Farming System. Soil Biology & Biochemistry, 23, 507-513.

42. Küsel, K. and Drake, H.L. (1999) Microbial Turnover of Low Molecular Weight Organic Acids during Leaf Litter Decomposition. Soil Biology & Biochemistry, 31, 107-118.

43. Calbrix, R., Barray, S., Chabrerie, O., Fourrie, L. and Laval, K. (2007) Impact of Organic Amendments on the Dynamics of Soil Microbial Biomass and Bacterial Communities in Cultivated Land. Applied Soil Ecology, 35, 511-522.

44. Pascual, J.A., Garcia, C., Hernandez, T. and Ayuso, M. (1997) Changes in the Microbial Activity of an Arid Soil Amended with Urban Organic Wastes. Biology and Fertility of Soils, 24, 429-434.

45. Kong, W.D., Zhu, Y.G., Fu, B.J., Han, X.Z., Zhang, L. and He, J.Z. (2008) Effect of Long-Term Application of Chemical Fertilizers on Microbial Biomass and Functional Diversity of a Black Soil. Pedosphere, 18, 801-808.

46. Li, J., Zhao, B.Q., Li, X.Y., Jiang, R.B. and Bing, S.H. (2008) Effects of Long-Term Combined Application of Organic and Mineral Fertilizers on Microbial Biomass, Soil Enzyme Activities and Soil Fertility. Agricultural Sciences in China, 7, 336-343.

47. Houot, S. and Chaussod, R. (1995) Impact of Agricultural Practices on the Size and Activity of the Microbial Biomass in a Long-Term Field Experiment. Biology and Fertility of Soils, 19, 309-316.

48. Ros, M., Garcia, C. and Hernandez, T. (2003) Soil Microbial Activity after Restoration of a Semiarid Soil by Organic Amendments. Soil Biology & Biochemistry, 35, 463-469.

49. Sánchez-Monedero, M.A., Mondini, C., Cayuela, M.L., Roig, A., Contin, M. and De Nobili, M. (2008) Fluorescein Diacetate Hydrolysis, Respiration and Microbial Biomass in Freshly Amended Soils. Biology and Fertility of Soils, 44, 885-890.

50. Dick, R.P. (1997) Soil Enzyme Activities as Integrative Indicators of Soil Health. In: Pankhurst, C.E., Doube, B.M. and Gupta, V.V.S.R., Eds., Biological Indicators of Soil Health, CAB International, Wallingford, 121-157.

51. Taylor, J.P., Wilson, B., Mills, M.S. and Burns, R.G. (2002) Comparison of Microbial Numbers and Enzymatic Activities in Surface Soils and Subsoils Using Various Techniques. Soil Biology & Biochemistry, 34, 387-401.

52. Jensen, L.E. and Nybroe, O. (1999) Nitrogen Availability to Pseudomonas fluorescens DF57 Is Limited during Decomposition of Barley Straw in Bulk Soil and in the Barley Rhizosphere. Applied and Environmental Microbiology, 65, 4320-4328.

53. Nannipieri, P.J., Ascher, M.T., Ceccherini, L., Landi Pietramellara, G. and Renella, G. (2003) Microbial Diversity and Soil Functions. European Journal of Soil Science, 54, 655-670.

54. Crecchio, C., Curci, M., Pizzigallo, M.D.R., Ricciuti, P. and Ruggiero, P. (2004) Effects of Municipal Solid Waste Compost Amendments on Soil Enzyme Activities and Bacterial Genetic Diversity. Soil Biology & Biochemistry, 36, 1595-1605.

55. Drissner, D., Blum, H., Tscherko, D. and Kandeler, E. (2007) Nine Years of Enriched CO2 Changes the Function and Structural Diversity of Soil Microorganisms in a Grassland. European Journal of Soil Science, 58, 260-269.

56. Bi, L., Zhang, B., Liu, G., Li, Z., Liu, Y., Ye, C., Yu, X., Lai, T., Zhang, J., Yin, J. and Liang, Y. (2009) Long-Term Effects of Organic Amendments on the Rice Yields for Double Rice Cropping Systems in Subtropical China. Agriculture, Ecosystems & Environment, 129, 534-541.

57. Singh, Y., Singh, B., Maskina, M.S. and Meelu, O.P. (1988) Effect of Organic Manures, Crop Residues and Green Manure (Sesbania aculeata) on Nitrogen and Phosphorus Transformations in a Sandy Loam at Field Capacity and under Waterlogged Conditions. Biology and Fertility of Soils, 6, 183-187.

58. Kaewpradit, W., Toomsan, B., Cadisch, G., Vityakon, P., Limpinuntana, V., Saenjan, P., Jogloy, S. and Patanothai, A. (2008) Regulating Mineral N Release by Mixing Groundnut Residues and Rice Straw under Field Conditions. European Journal of Soil Science, 59, 640-652.

59. Gaur, A.C. and Pareek, R.P. (1974) Organic Acids in Soil during Degradation of Organic Residues. Proceedings of the Indian National Science Academy, Part B, 40, 68-76.

60. Nelson, C.J. (1996) Allelopathy in Cropping Systems. Agronomy Journal, 88, 853.

61. Azam, F., Lodhi, A. and Ashraf, M. (1991) Availability of Soil and Fertilizer Nitrogen to Wetland Rice Following Wheat Straw Amendment. Biology and Fertility of Soils, 11, 97-100.

62. Verma, T.S. and Bhagat, R.M. (1992) Impact of Rice Straw Management Practices on Yield, Nitrogen Uptake and Soil Properties in a Wheat-Rice Rotation in Northern India. Fertilizer Research, 33, 97-106.

CHAPTER 8

The Effect of the Geomorphologic Type as Surrogate to the Time Factor on Digital Soil Mapping

Huixia Chai[1], Sheng Rao[1], Reibo Wang[2], Jing Liu[3], Qi Huang[1], Xuejie Mou[1]

[1]Chinese Academy for Environmental Planning, MEP, Beijing, China

[2]State Key Laboratory of Resources and Environmental Information System, Institute of Geographical Sciences and Natural Resources Research, CAS, Beijing, China

[3]Department of Geography, University of Wisconsin-Madison, Madison, USA

ABSTRACT

Many environmental variables are frequently used to predict values of soil in locations where they are not measured. Digital soil mapping (DSM) has a long-standing convention to describe soils as a function of climate, organisms, topography, parent material, time and space. It is obvious that terrain, climate, parent material and organisms are used frequently in the prediction of soil properties while time and space factors are rarely used. Time is the indirect factor for the formation and development of soil. Moreover, it is very useful to explicit and implicit estimates of soil age for DSM. However, it is often difficult to obtain time factor. In the absence of explicit soil age data, geomorphologic data are commonly related to soil relative age. Consequently, this study adopts the geomorphologic types (genesis type of geomorphology) as surrogate to the time factor and analyzes its effect on DSM. To examine this idea, we selected the Ili region of northwestern China as the study area. This paper uses geomorphologic data from a new digital geomorphology map as the implicit soil age in predictive soil mapping. For this study, Soil-landscape inference model (SoLIM) was used to predict soil properties based on the individual representation of each sample. This model applies the terrain (topography), climate, parent material (geology) and time (geomorphologic type) to predict soil values in the study area where they are not measured. And the independent sample validation method was used to estimate the precision of results. The validation result shows that the use of geomorphologic data as surrogate to the time factor in the individual representation leads to a considerable and significant increase in the accuracy of results. In other words, implicit estimates of soil age by genesis type of geomorphology are very useful for DSM. This increase was due to the high purity of the geomorphologic data. This means that the geomorphologic variable, if used, can improve the quality of DSM. Predicted value through the proposed approach comes closer to the real value.

KEYWORDS

Geomorphologic Type, Environmental Variables, Digital Soil Mapping, Soil-Landscape Inference Model (SoLIM), Individual Validation

1. INTRODUCTION

Digital soil mapping (DSM) is a technology and method to describe the spatial distribution information of soil with a grid format on the basis of field samples and/or auxiliary data, such as soil environmental factors and other soil properties [1] . DSM has a long-standing convention to describe soils as an equation of climate, organisms, topography, parent material, and time [2] [3] . McBratney et al. [4] put forward a so-called "scorpan model" as a spatial predicted model of soil properties based on the environmental variables of climate, organisms, topography, parent material, time and space. Those variables, if their information is available for the location, are frequently used to predict values of soil in locations where they are not measured [5] - [9] . Therefore, these factors should consider into the methods or models of predicted soil properties, at least in theory. But in practice it is hard to do because it is too difficult to get or measure some environmental factors like time.

Studies use various approaches to predict soil properties by single-variable and multi-variable statistics [10] - [12] , geostatistical [13] - [15] and hybrid methods [16] [17] , and process-based models that relate soils to environmental covariates consider spatial and temporal dimensions [1] [4] [18] . Recently, most studies adopt the synthetic analysis and comprehensive applications of more environmental variables to predict the soil properties or the spatial distribution of soil [19] - [22] . However, it is obvious that terrain, climate, parent material (geology) and organisms (vegetation) are used frequently in the prediction of soil properties while time and space factors are rarely used.

Actually, it is very necessary to get the time factor into DSM. Time is the indirect factor for the formation and development of soil because it is a way to embody the effect of climate, organisms, topography, and parent material on the soil formation through time [3] . Explicit and implicit estimates of soil age are very useful environmental covariants for DSM, but are often difficult to obtain [12] . The temporal information of soil formation is unavailable, or not widely available, especially the absolute time, although some studies use the 14C detection or fossil analysis to obtain the soil age of sample points [23] [24] . Moonjun et al. [25] tried to find a correlation between the observed data (in the sample areas) and the environmental characteristics derived from the DEM, and they distinguish three relative soil ages: old (Ultisols and Oxisols), mature (Alfisols), and young (Inceptisols and Entisols). Noller [26] reviews various approaches to add geochronology in DSM and obtains the improved prediction of soil on the landscape in addition to several implicit age covariates, including custom thematic maps of lithology and geologic age. Therefore, it is important to get the time factor into DSM. Furthermore, how to accurately get the temporal information of soil formation is always a problem in the whole study area only based on few observed data of field samples. Generally, in the absence of explicit soil age data, implicit topographic data representing landforms that are commonly related to soil relative age and remotely sensed spectral data that focus on surface spectral characteristics represent surface or soil age [12] .

Consequently, the purpose of this study is to analyze the effect of the genesis types of geomorphology as surrogate to the time factor on DSM. In other words, this study adopts the geomorphologic types as the relative time in the process of DSM. To analyze the effect of geomorphologic types on DSM, this study uses the SoLIM (a professional software for DSM) to predict soil organic matter (%) of the top layer in locations where they are unknown.

2. BASIC IDEA

2.1. Theory

Geomorphology is one of the predominant factors determining the features of the earth's surface and an important symbol and dominant factor of regional differentiation. It (geomorphology) directly influences and even decides the spatial distribution and transformation of hydrology, climate, soil, and other ecological and environmental factors [27] . Since in 1980, Knott et al. [28] had been analyze and find that the geomorphologic mapping boundaries are coincident with soil material boundaries, which implies a correlation between soil material and geomorphology to some degree. Geomorphologic processes, both erosional and depositional, create distinctive landforms which have a great influence on soil types and soil distribution [29] . The soil condition is different in places with various geomorphologic types due to the forming reason of geomorphology, as well as in different parts of geomorphology. The forming reason of geomorphology means that it will appear in some dominant geomorphologic processes, and produce characteristic landforms in particular climatic conditions [30] . Bacon et al. [31] point that systematic relationships exist between landscapes position (terrain conditions of desert landform) and soil formation processes. In their study, the assignment of relative age classes to each landform or geomorphic surface map unit was based on crosscutting relations, surface morphology and roughness, and topographic relief observable on the multispectral imagery. Moreover, Stum et al. [32] use landform surface age as the implicit soil age in predictive soil mapping by spatially modeling the shoreline of Pleistocene pluvial Lake Bonneville.

Therefore, we consider that the geomorphologic type could be used to approximate the time factor at the categorical level in soil formation processes. In other words, this study wants to take the geomorphologic type as surrogate to the time factor on DSM. The geomorphologic type here specifically means the genesis type of geomorphology. The genesis type is classified by the forming reason of geomorphology. It is a nominal variable and only represents the relative time.

2.2. Experiment Design

The process of DSM in this study was based on the Soil-landscape inference model (SoLIM). SoLIM is a new technology for soil mapping based on recent developments in Geographic Information Sciences (GIS), Artificial Intelligence, and Information Representation Theory. SoLIM was designed to overcome the limitations of existing soil survey methods and to improve the efficiency and accuracy of soil surveys [3] . Case studies have shown that SoLIM is more efficient and accurate than traditional soil survey methods, in that it generates a range of products which the traditional approaches couldn't provide, and it can be employed in a production mode of soil survey.

This study adds the geomorphologic type as surrogate to the relative time of soil and use the sampled-based approach under the SoLIM framework to examine the effect of including geomorphologic type. The environmental variables in this study (Table 1) including topography (Terrain, derive from DEM), climate, parent material (obtain from geology map) and time (relative time, using geomorphologic type as indicator).

This study describes two experiments to test the idea mentioned above. Each experiment has different environmental variables. The first group of environmental variables includes climate, parent material and topography. The second group includes climate, parent material, topography and geomorphology. Based on two groups of environmental variables, this paper presents two soil maps by the same research methods. Then we compare and analyze the precision between the two soil maps.

3. METHOD AND DATA

This section provided an introduction to study area, data and method in this study. By the way, readers can refer to literatures [3] [33] if you are interested in a comprehensive description of prediction method. Based on the knowledge of the relationship between soil and environment, this paper selected four environmental (Table 1) variables to describe environmental features. Climate data and geomorphic types collected by State Key Laboratory of Resource and Environmental Information System, IGSNRR, CAS. Parent material rasterized from Chinese Geology Map (1:1000,000). Seven topographic covariates (altitude (m), slope gradient (%), aspect, contour curvature, profile curvature, Plane curvature, Surface area ratio (SAR), Topographical wetness index (TWI)) were derived from DEM, which was downloaded from SRTM website.

3.1. Study Area

The study area of this research is the Ili region located between 42°14'16"N and 44°50'30"N, 80°09'42"E and 84°56'50"E, Xinjiang Uygur Autonomous Region, China (Figure 1). The Ili region is about 350 km (east-west) by 280 km (north-south) covering about 56,400 km^2. Terrain conditions here are complex, containing arduous mountains as well as flat plains. Influenced by local geomorphology, climate and vegetation conditions deviate from the general distribution pattern at certain locations. The amount of annual precipitation on the flat plain is about 140 - 500 mm and, as elevation increases from west to east, annual precipitation increases at the same time. Vegetation types mainly include meadows and swamps on flat plains and forests on mountains.

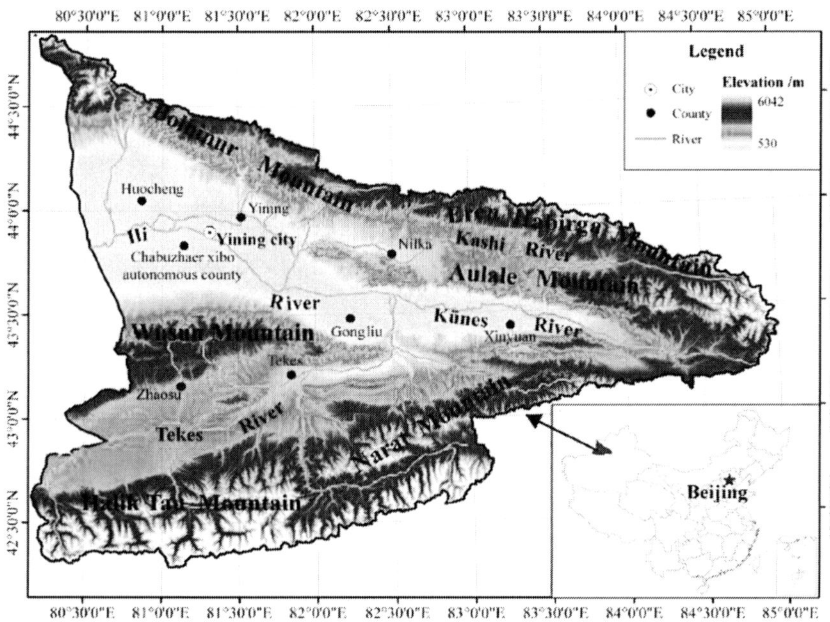

Figure 1. Study area.

The main parent materials are quartz sand, mica sand, red mudstones and alluvial loess or alluvial loess-like materials. Thanks to the variation of climate, terrain, vegetation and parent material conditions, soil types are relatively abundant on both vertical and horizontal distributions, mainly including chestnut soil, chernozem soil, forest soil, meadow soil, and Sierozem soil.

The basic geomorphology in the study area is composed of mountain landform and river-valley landforms. Geomorphologic types in the mountains at extremely high or high altitudes are mainly glacial and periglacial landforms. The lower part of the mountains is mainly fluvial landforms which formed by the erosion action of water and arid landforms which formed by a dry eroding effect. Alluvial plain, proluvial plain, and alluvial-pro- luvial plain are mainly distributed in the basin area. Fluvial landform is the dominant geomorphologic type in the plain landform area. In addition, the study area also has a small number of aeolian landform and loess landform.

Table 1. Environmental variables and their resolution.

Environmental variables		Resolution	Data sources
Climate	Precipitation Temperature Relative humility	1 km	State Key Laboratory of Resource and Environmental Information System, IGSNRR, CAS
Parent material	Parent material type	1 km	Rasterized from Chinese Geology Map (1:1000,000)
Topography	Altitude Slope gradient Aspect Profile curvature Plane curvature Surface area ratio, SAR Topographical wetness index, TWI	90 m	Original DEM was downloaded from SRTM website and derived indexes were generated based on either ArcGIS or specific algorithms
Geomorphology	Geomorphologic type (genesis type of geomorphology)	1 km	State Key Laboratory of Resources and Environmental Information System, IGSNRR, CAS

3.2. Geomorphologic Data and Field Data

This paper collected all necessary data of study area, including climate data, SRTM-DEM, Geology Map, and the data of geomorphic types, to gain corresponding variables.

Geomorphology has a profound influence on the formation and development of soil by redistributing parent material, water, and heat. The geomorphologic type derives from the new 1:1,000,000 set of geomorphologic atlas of China. The digital geomorphologic database was created by visual interpretation from Landsat TM/ETM imageries, SRTM-DEM, and published geomorphologic maps, etc. [34] . The atlas was compiled and finished from the database, published by Science Press in 2009 [35] . The geomorphologic data include seven layers, i.e. basic morphology, genesis, sub-genesis, morphology, micro-morphology, slope and aspect, material composition and lithology. Here, we only use the genesis type of geomorphology to represents the relative time.

In addition, this paper also collected 107 soil sample data from the following sources: 1) National Native Soil Type Log (the Second National Soil Survey Office, 1995); 2) Field survey carried out by Dr. Ma Xingwang, Institute of Soil and Fertilizer, Xinjiang Academy of Agricultural Sciences; 3) Offered by Dr. Zhang Baiping and Dr. Zhang Hongqi, Institute of Geographical Sciences and Natural Resources Research, Chinese Academy of Sciences; 4) The book named Soil in Ili area, Xinjiang (1985).

3.3. Prediction Method and Computational Details

This paper adopts the independent sample validation method to estimate the performance of the predicted result. First, 75 of the 107 sampling were chosen to serve as sample points to predict soil properties. Then the other 32 soil samples serve as validation points to test the precision of DSM. Finally, estimate value of validation points would be extracted from the predicted result of 75 sample points through add validation points layer upon the result layer. This method can examine the precision of single point forecasting on two soil maps.

3.3.1. Environmental Variables

Climate data include annual-averaged precipitation, temperature and relative humidity, as well as the maximum and minimum of monthly-averaged precipitation, temperature and relative humidity within a year. These three variables have great influence on the decomposition and transformation of soil organic matter.

Topographical indexes include slope, aspect, profile curvature, plane curvature, surface area ratio and topographical wetness index. Topographical Wetness Index (TWI) is estimated according to a modified Multiple- Flow-Direction (MFD) algorithm [36] for calculating upslope drainage area. Surface Area Ratio (SAR), which represents the roughness within a mapping grid, is generated based on the algorithm developed by Jenness [37] . Other topographical variables are generated through ArcGIS. The content of surface soil organic matter often has a negative correlation to elevation [38] [39] . Slope, which represents the tilt degree of the local surface topography, has a direct relationship to the stability of soil, the drainage and gathering ability of surface water [39] [40] . Aspect affects the accumulation and decomposition of surface soil organic matter [41] . Curvature affects the spatial distribution of surface soil organic matter [42] . Surface Area Ratio (SAR) is used for distinguishing mountains and plains by terrain characteristics [43] . Topographical wetness index, which represents the drainage and gathering ability of soil, affects the content of surface soil organic matter [39] .

Parent material types are often used to describe the different characteristics of parent material [4] . They include quartz sand, mica sand, red mudstones, alluvial loess, alluvial loess-like materials, glacier and snow cover.

The geomorphologic type which mentioned in this paper is the genesis type of geomorphology. This study adopts the genesis type of geomorphology as surrogate to the time factor. In other words, the geomorphologic type should be regard as the relative time for DSM in this study. According to the forming reason of geomorphology, there are six geomorphologic types in the study area: loess landform, fluvial landform, Aeolian landform, arid landform, glacial landform, and periglacial landform.

3.3.2. Quantifying and Estimating Similarity

His study, as the resolution of climate, parent material and geomorphology information is approximate to the mapping grid, the single value method was adopted to represent climate and geological characteristics of each grid. As the resolution of topographical factors is refined, the probability distribution of each index was approximated to represent topographical characteristics of each grid. For estimating a smooth probability distribution curve, the Kernel Density Estimation (KDE) method has been adopted in this case study.

In this study GOWER distance was adopted to calculate climate similarity. GOWER distance measures the similarity between a given location (i, j) and a sample k on a series of environmental variables, standardized by the range of each variable [44] (Equation (1)):

$$Clim\,ate\_simi_{i,j}^{k} = \frac{1}{m}\sum_{v=1}^{m}\left(1-\frac{\left|e_{ij}^{v}-e_{k}^{v}\right|}{range\left(e^{v}\right)}\right) \tag{1}$$

where m is the number of climate variables; $\left|e_{ij}^{v}-e_{k}^{v}\right|$ denotes the absolute difference on the v-th climate variable between location (i, j) and sample k; $\text{range}\left(e^{v}\right)$ is the global value-range of the v-th climate variable. The value of overall climate similarity $Clim\,ate\_simi_{ij}^{k}$ will be between 0 and 1.

Parent material similarity and geomorphologic similarity were decided through Boolean function. If the parent material type and geomorphologic type at the location (i, j) are identical with sample point k, it means parent material similarity and geomorphologic similarity are equal to 1, conversely, parent material similarity and geomorphologic similarity equal 0.

Similarity on each topographical index is calculated through the Consistency Measure (CM) algorithm, and finally, the integrated topographical similarity for all topographical indexes was estimated by an average-ope- rator (Equation (2)).

$$Topographic\_simi_{i,j}^{k} = \frac{1}{m}\sum_{v=1}^{m} CM_{ij,k}^{v} \tag{2}$$

where $Topographic\_simi_{i,j}^{k}$ denotes the topographical similarity between location (i, j) and sample k; m is the number of involved topographical indexes; $CM_{ij,k}^{v}$ is the Consistency index of the probability density distribution function on the v-th topographical variable between location (i, j) and sample k.

Based on the importance of each environmental variable which indicates spatial distribution of soil at different scales, this paper generates the similarity between location point and sample point through comprehensive analysis of climate similarity, parent material similarity, topographical similarity and similarity of geomorphologic type (Figure 2 and Figure 3).

Please note the similarities and differences between Figure 2 and Figure 3.

Bothe figures have three things in common. Firstly, C_similarity, Topo_similarity means climate similarity, topographical similarity separately. Secondly, weights of climate similarity and topographical similarity both are 0.5. Thirdly, climate similarity cut is 0.5.

Figure 2 and Figure 3 have one difference. In Figure 2, environmental variables include climate, parent material and topography. Figure 3 has one more environmental variable than Figure 2, environmental variables include climate, parent material, topography and geomorphologic type.

First, this paper examines the climate similarity between an unsampled location and a certain case (a climate- similarity-cut was set at 0.5 and has been adopted generally as we had no background information on deciding this threshold).

Then, we test the parent material similarity if the unsampled location was similar enough with the case on climate.

Third, this paper continue to examine the geomorphologic similarity if the parent material of the unsampled location is the same as the case (it means the parent material similarity is equal to 1).

Finally, the topographical similarity and climate similarity would be integrated to estimate the final similarity using the weighted-average method if the geomorphologic similarity is equal to 1 (it means the geomorphologic type of the unsampled location is the same as the case).

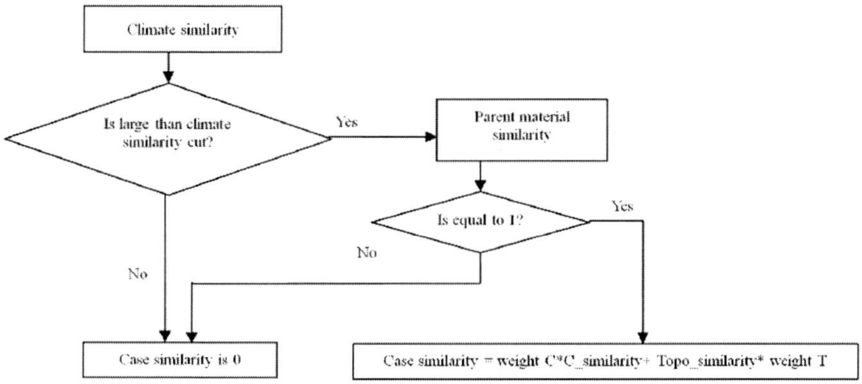

Figure 2. P-function for integrating environmental similarity into case-similarity.

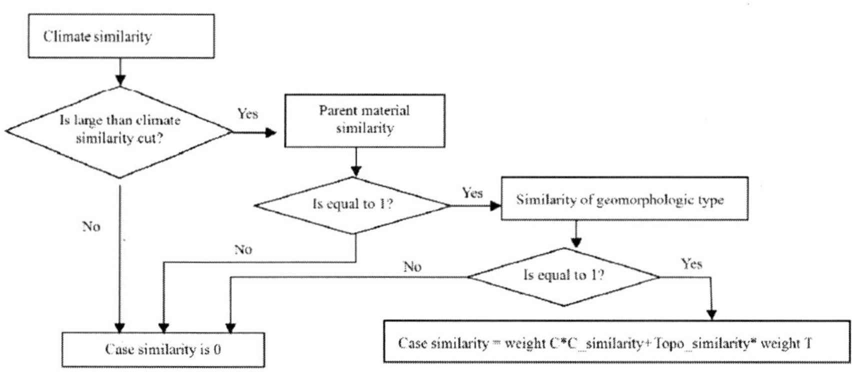

Figure 3. P-function for integrating environmental similarity into case-simi- larity.

Here, the weights of the topographical similarity and climate similarity should be assigned considering the relative importance between them. As we had no knowledge to decide which one is more important, we assigned 0.5 to both as the weights.

The estimation of the value of the targeted soil property is calculated with a weighted average method [3]:

$$V_{ij} = \frac{\sum_{k=1}^{n} S_{ij}^{k} \cdot V^{k}}{\sum_{k=1}^{n} S_{ij}^{k}} \qquad (3)$$

in which n' is the number of the selected soil samples whose environmental similarity to the unvisited location (i, j) exceeds 1 minus the uncertainty threshold, S_{ij}^{k} is the environmental similarity of the unvisited location (i, j) to the soil type k in sample location k, and V^k is the value of the targeted soil property of soil sample location k.

4. RESULT AND DISCUSSION

4.1. Predicted and Discussion

Figure 4 shows soil organic matter (%) of the top layer predicted through the proposed approaches. The gray area with "No-Data" is the grid which cannot be represented by current samples. Figure 4(b) is the result on the basis of three environmental variables: climate, parent material and topography. Figure 4(b) is the result on the basis of four environmental variables: climate, parent material, topography and geomorphology.

The result shows that estimated value through the proposed approaches is closer to the real value than to the average value. In other words, the proposed approach in this paper is valid. These graphs (Figure 4) show that the spatial distribution of soil organic matter content has little difference between the two results. It can help avoid wasting time and energy if the geomorphologic type is added as surrogate to the time factor in the process of DSM. This means it will improve work efficiency.

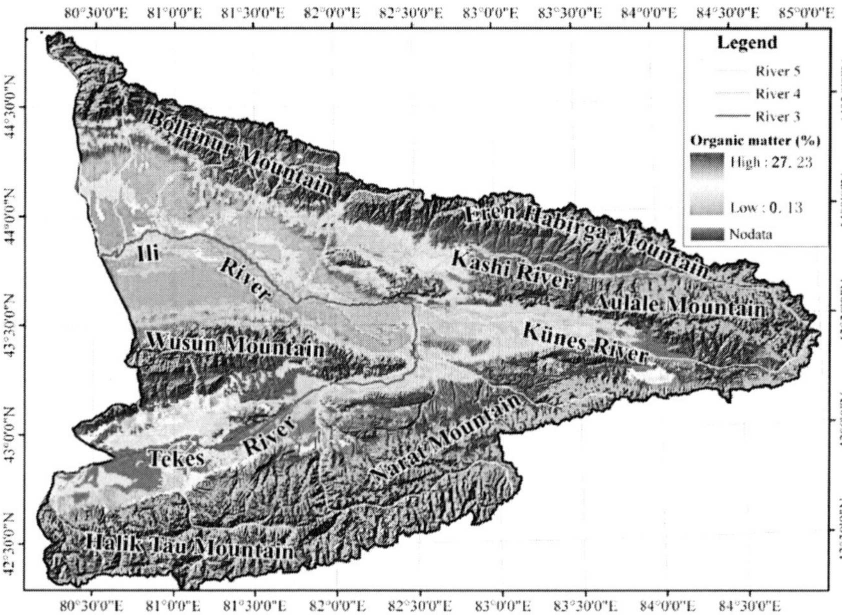

(a)

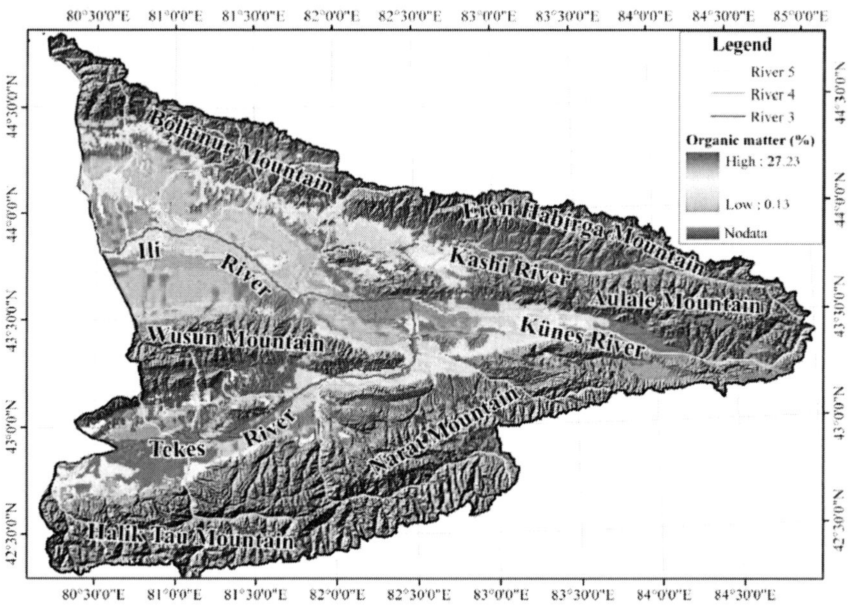

(b)

Figure 4. Soil organic matter (%) of top layer predicted though the proposed approach. (a) Without geomorphology; (b) With geomorphology.

4.2. Validation and Discussion

Generally, the mean/mean-bias error (ME or MBE), the mean absolute error (MAE), and the root mean squared error (RMSE) are used to test the error statistic. In this paper, the deviations of validation results are summarize by RMSE and MAE.

The root mean squared error (RMSE) provide a measure of estimation error that was more sensitive to outliers than the MAE

$$RMSE = \sqrt{\frac{1}{n}\sum_{i=1}^{n}\left[V'(s_i) - V(s_i)\right]^2} \quad (3)$$

where the generated predicted value represent by $V'(s_i)$, the true observation value represent by $V(s_i)$. So, predicted error at each sample point describe by $V'(s_i) - V(s_i)$.

The mean absolute error (MAE) reflects the absolute error of the estimates. It is expressed as follows

$$MAE = \frac{1}{n}\sum_{i=1}^{n}\left|V'(s_i) - V(s_i)\right| \quad (4)$$

For 32 validation points, only 20 points have effective values, the rest of the validation points are located in the place where without predicted value. So, this paper just adopts the predicted values of these 20 points to calculate RMSE and MAE.

Table 2 shows the validation results of two experiments within different environmental variables.Table 2 indicates that the root mean squared error (RMSE) and the mean absolute error (MAE) are lower significantly after the data layer of geomorphologic type is added. It explains that the geomorphology variable can improve the accuracy of the predicted results in soil organic matter mapping. The values of RMSE and MAE with geomorphology are less than that without geomorphology. The value of RMSE without geomorphology is three times as big as the values of RMSE with geomorphology. This result indicating that the accuracy of predicted results with geomorphologic type is higher than one without geomorphologic type.

According to the validation results, most predicted values with geomorphology of validation samples are closer to real values than predicted values without geomorphology (Figure 5 and Figure 6). It can veritably reflect the soil properties if the geomorphologic type is added as the relative time. After adding the geomorphology variable, the accuracy rate of predicted values have been increased by an average of 34%. In other words, the predicted result with geomorphology is much better than the predicted result without geomorphology.

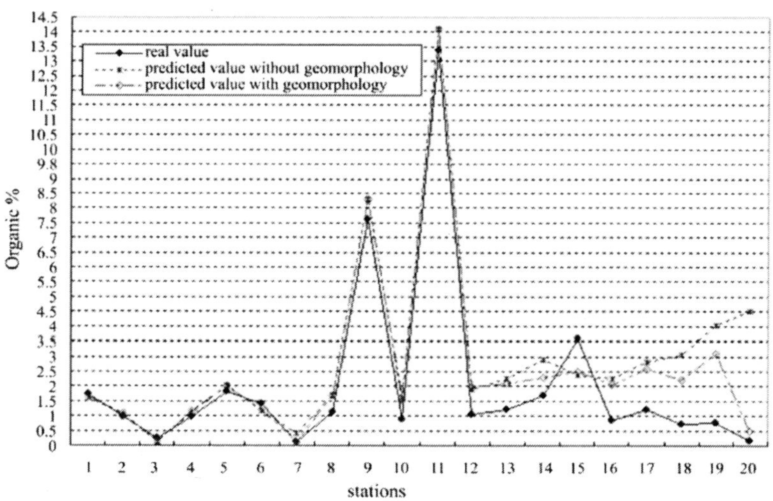

Figure 5. Comparative analysis of the real values, predicted values without geomor- phology and with geomorphology.

Table 2. Validation results.

	Without geomorphology	With geomorphology
RMSE	2.33	0.81
MAE	1.06	0.69

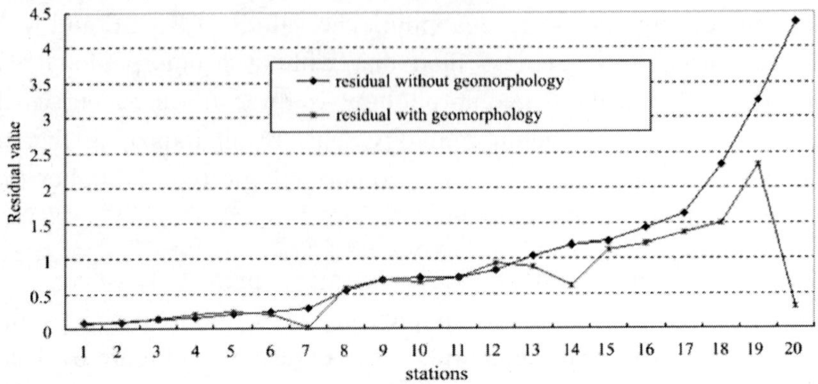

Figure 6. Comparative residual of the predicted values without geomorphology and the predicted values with geomorphology.

5. CONCLUSIONS

There is no universal soil equation or digital soil prediction model that fits all regions and purposes [11] . Hence, the soil prediction at different scales (local, national and global scales) needs to build various equations and models respectively. This study adopts the geomorphologic type to represent the relative time of soil age and estimate the effect of the geomorphologic type on the predicted precision of DSM. We consider genesis type of geomorphology as surrogate to the time factor and predict soil organic matter through the SoLIM. This study tries to analyze the effect of the geomorphologic types as surrogate to the time factor on DSM by the case study.

The result shows that using the geomorphologic type as surrogate to the time factor can improve the result quality of DSM. The genesis type of geomorphology results in the estimated value closer to the real value. Validation with a probability sample shows that use of geomorphology data as the relative time factor leads to a considerable and significant increase of accuracy. In other words, implicit estimates of soil age by genesis type of geomorphology are very useful for DSM. There are 75% predicted values having been increased. This increase in accuracy is due to the high precision of the existing geomorphology data.

In addition, with such a limited sample-set and such complex natural conditions, it is probably not reasonable to adopt any of the current interpolation method to predict soil property values over the entire study area. That's why there are many empty values in Figure 4. Therefore, in order to further test the degree of influence of the geomorphology variable on the precision of DSM, more samples are needed. This will be the author's next step, a quantitative description of the degree of influences of the geomorphology variable on the precision of DSM.

ACKNOWLEDGEMENTS

The authors would like to thank anonymous reviewers for very helpful suggestions that were used to improve the manuscript.

REFERENCES

1. McBratney, A.B., Odeh, I.O.A., Biship, T., Dunbar, M.S. and Shatar, T.M. (2000) An Overview of Pedometric Techniques for Use in Soil Survey. Geoderma, 97, 293-327.

2. Jenny, H. (1941) Factors of Soil Formation: A System of Quantitative Pedology. Dover Publications, New York, 281 p. http://books.google.com/books?id=p94TjEnzCTcC&dq=Factors+of+Soil+Formation:+A+System+of+Quantitative+Pedology&printsec=frontcover&source=bn&hl=en&ei=XfXZS_udG5OINtr74Ug&sa=X&oi=book_result&ct=result&resnum=4&ved=0CBkQ6AEwAw#v=onepage&q&f=false

3. Zhu, A.X., Li, B.L. and Pei, T. (2008) Model and Method of Detail Digital Soil Survey. Science Press, Beijing. (In Chinese)

4. McBratney, A.B., Mendonça Santos, M. and Minasny, B. (2003) On Digital Soil Mapping. Geoderma, 117, 3-52.

5. Zhu, H.J. and He, Y.G. (1991) Soil Geography. Higher Education Press, Beijing. (In Chinese)

6. Wu, G.H., Tian, L.S., Hu, S.X. and Wang, N.A. (2000) Natural Geography. Higher Education Press, Beijing. (In Chinese)

7. Florinsky, I.V., Eilers, R.G., Manning, G.R. and Fuller, L.G. (2002) Prediction of Soil Properties by Digital Terrain Modelling. Environmental Modelling & Software, 17, 295-311.

8. Li, T.J., Zhao, Y., Zhang, K.L., Zheng, Y.S. and Wang, Y. (2004) Soil Geography. Higher Education Press, Beijing. (In Chinese)

9. Mendonça-Santos, M.L., Dart, R.O., Santos, H.G., Coelho, M.R., Berbara, R.L.L. and Lumbreras, J.F. (2010) Digital Soil Mapping of Topsoil Organic Carbon Content of Rio de Janeiro State, Brazil. In: Boettinger, J.L., Howell, D.W., Moore, A.C., Hartemink, A.E. and Kienast-Brown, S., Eds., Digital Soil Mapping, Springer Science + Business Media B.V., New York, 255-266.

10. Moore, I.D., Gessler, P.E., Nielsen, G.A. and Peterson, G.A. (1993) Soil Attributes Prediction Using Terrain Analysis. Soil Science Society of America Journal, 57, 443-452.

11. Grunwald, S. (2009) Multi-Criteria Characterization of Recent Digital Soil Mapping and Modeling Approaches. Geoderma, 152, 195-207.

12. Boettinger, J.L. (2010) Environmental Covariates for Digital Soil Mapping in the Western USA. In: Boettinger, J.L., Howell, D.W., Moore, A.C., Hartemink, A.E. and Kienast-Brown, S., Eds., Digital Soil Mapping, Springer Science + Business Media B.V., New York, 17-27.

13. Odeh, I.O.A., McBratney, A.B. and Chittleborough, D.J. (1992) Fuzzy-c-Means and Kriging for Mapping Soil as a Continuous System. Soil Science Society of America Journal, 56, 1848-1854.

14. Bourgault, G., Journel, A.G., Rhoades, J.D., Corwin, D.L. and Lesch, S.M. (1997) Geostatistical Analysis of a Soil Data Set. Advances in Agronomy, 58, 241-292.

15. Hendricks Frassen, H.J.W.M., van Eijnsbergen, A.C. and Stien, A. (1997) Use of Spatial Prediction Techniques and Fuzzy Classification for Mapping Soil Pollutants. Geoderma, 77, 243-262.

16. Odeh, I.O.A., McBratney, A.B. and Chittleborough, D.J. (1995) Further Results on Prediction of Soil Properties from Terrain Attributes: Heterotopic Cokriging and Regression-Kriging. Journal of Geoderma, 67, 215-225.

17. Odeh, I.O.A. and McBratney, A.B. (1998) Using NOAA Advanced Very High Resolution Radiometric Imageries for Regional Soil Inventory. Proceedings of the 16th World Congress of Soil Science, Montpellier, 20-26 August 1998.http://natres.psu.ac.th/Link/SoilCongress/bdd/symp17/498-r.pdf

18. Grunwald, S. (2010) Current State of Digital Soil Mapping and What Is Next. In: Boettinger, J.L., Howell, D.W., Moore, A.C., Hartemink, A.E. and Kienast-Brown, S., Eds., Digital Soil Mapping, Springer Science + Business Media B.V., New York, 3-12.

19. Boettinger, J.L., Ramsey, R.D., Bodily, J.M., Cole, N.J., Kienast-Brown, S., Nield, S.J., Saunders, A.M. and Stum, A.K. (2008) Landsat Spectral Data for Digital Soil Mapping. In: Hartemink, A.E., McBratney, A.B. and MendonÇa-Santos, M.L., Eds., Digital Soil Mapping with Limited Data, Springer Science + Business Media B.V., New York, 193-202.

20. Gonzalez, J.P., Jarvis, A., Cook, S.E., Oberthür, T., Rincon-Romero, M., Bagnell, J.A. and Dias, M.B. (2008) Digital Soil Mapping of Soil Properties in Honduras Using Readily Available Biophysical Datasets and Gaussian Processes. In: Hartemink, A.E., McBratney, A.B. and MendonÇa-Santos, M.L., Eds., Digital Soil Mapping with Limited Data, Springer Science + Business Media B.V., New York, 367-380.

21. Roecker, S.M., Howell, D.W., Haydu-Houdeshell, C.A. and Blinn, C. (2010) A Qualitative Comparison of Conventional Soil Survey and Digital Soil Mapping Approaches. In: Boettinger, J.L., Howell, D.W., Moore, A.C., Hartemink, A.E. and Kienast-Brown, S., Eds., Digital Soil Mapping, Springer Science + Business Media B.V., New York, 369-384.

22. Wiesmeier, M., Barthold, F., Blank, B. and Kögel-Knabner, I. (2011) Digital Mapping of Soil Organic Matter Stocks Using Random Forest Modeling in a Semi-Arid Steppe Ecosystem. Plant and Soil, 340, 7-24.

23. Wysocki, D.A., Schoeneberger, P.J. and LaGarry, H.E. (2000) Geomorphology of Soil Landscapes. In: Sumner, M.E., Ed., Handbook of Soil Science, CRC Press, New York, E5-E39.

24. Yang, M.S., Zhang, H.C., Ding, H., Lei, G.L., Fan, H.F., Chang, F.Q., Li, B., Zhang, W.X., Niu, J. and Chen, Y. (2005) 14C Dating of Paleosol and Animal Remains in Loess Deposit: A Comparative Study. Journal of China University of Geosciences, 30, 589-596.

25. Moonjun, R., Farshad, A., Shrestha, D.P. and Vaiphasa, C. (2010) Artificial Neural Network and Decision Tree in Predictive Soil Mapping of Hoi Num Rin Sub-Watershed, Thailand. In: Boettinger, J.L., Howell, D.W., Moore, A.C., Hartemink, A.E. and Kienast-Brown, S., Eds., Digital Soil Mapping, Springer Science + Business Media B.V., New York, 151-164.

26. Stratton, N.J. (2010) Applying Geochronology in Predictive Digital Mapping of Soils. In: Boettinger, J.L., Howell, D.W., Moore, A.C., Hartemink, A.E. and Kienast-Brown, S., Eds., Digital Soil Mapping, Springer Science + Business Media B.V., New York, 43-53.

27. Chai, H.X., Zhou, C.H., Chen, X., Cheng, W.M., Ou, Y. and Yuan, Y.C. (2008) The New Methodology of Geomorphological Zonalization in Xinjiang Based on Geographical Grid. Journal of Geophysical Research, 27, 481-492. (In Chinese)

28. Knott, P.A., Doornkamp, J.C. and Jones, R.H. (1980) The Relationship between Soils and Geomorphologic Mapping Units—A Case Study from Northamptonshire. Bulletin of the International Association of Engineering Geology, 21, 186-193.

29. Gerrard, J. (1992) Soil Geomorphology: An Integration of Pedology and Geomorphology. Chapman and Hall, London.

30. Zhou, C.H. (2006) A Dictionary of Geomorphology. China Water Power Press, Beijing. (In Chinese)

31. Bacon, S.N., McDonald, E.V., Dalldorf, G.K., Baker, S.E., Sabol Jr., D.E., Minor, T.B., Bassett, S.D., MacCabe, S.R. and Bullard, T.F. (2010) Predictive Soil Maps Based on Geomorphic Mapping, Remote Sensing, and Soil Databases in the Desert Southwest. In: Boettinger, J.L., Howell, D.W., Moore, A.C., Hartemink, A.E. and Kienast-Brown, S., Eds., Digital Soil Mapping, Springer Science + Business Media B.V., New York, 411-421.

32. Stum, A.K., Boettinger, J.L., White, M.A. and Ramsey, R.D. (2010) Random Forests Applied as a Soil Spatial Predictive Model in Arid Utah. In: Boettinger, J.L., Howell, D.W., Moore, A.C., Hartemink, A.E. and Kienast-Brown, S., Eds., Digital Soil Mapping, Springer Science + Business Media B.V., New York, 179-190.

33. Liu, J. (2010) Mapping Soil Properties Using Individual Representativeness of Samples over Large Areas. Master's Dissertation, Beijing Normal University, Beijing.

34. Cheng, W.M., Zhou, C.H., Chai, H.X., Zhao, S.M., Liu, H.J. and Zhou, Z.P. (2011) Research and Compilation of the Geomorphologic Atlas of the People's Republic of China (1:1,000,000). Journal of Geographical Sciences, 21, 89-100.

35. Li, J.J., Zhou, C.H. and Cheng, W.M. (2009) Geomorphologic Atlas of the People's Republic of China. Science Press, Beijing. (In Chinese)

36. Qin, C.Z., Zhu, A.X., Pei, T., Li, B.L., Zhou, C.H. and Yang, L. (2007) An Adaptive Approach to Selecting a Flow- Partition Exponent for a Multiple-Flow-Direction Algorithm. International Journal of Geographical Information Science, 21, 443-458.

37. Jenness, J.S. (2004) Calculating Landscape Surface Area from Digital Elevation Models. Wildlife Society Bulletin, 32, 829-839.

38. Qiu, Y., Fu, B.J., Wang, J. and Chen, L.D. (2004) Spatio-Temporal Variation and Its Influencing Factors of Soil Nutrients in a Small Catchment of the Loess Plateau. Journal of Progress in Natural Science, 14, 294-299. (In Chinese)

39. Lian, G., Guo, X.D., Fu, B.J. and Hu, C.X. (2006) Spatial Variability and Prediction of Soil Organic Matter at County Scale on the Loess Plateau. Progress in Geography, 25, 112-122. (In Chinese)

40. Huang, P., Li, T.X., Zhang, J.B. and Liao, G.T. (2009) Effects of Slope's Gradient and Aspect on Spatial Heterogeneity of Soil Organic Matter in Low Mountainous Region. Journal of Soils, 41, 264-268. (In Chinese)

41. Zhong, Y.P., Tang, J. and Wang, L. (2006) Distribution Characteristic of Soil Organic Carbon in Three Gorges Reservoir District. Journal of Soil and Water Conservation, 20, 73-76. (In Chinese)

42. Zhou, B., Xu, H.W. and Wang, R.C. (2003) Soil Organic Matter Mapping Based on Classification Tree Modeling. Acta Pedologica Sinica, 40, 801-808. (In Chinese)

43. Jenness, J. (2009) Surface Area and Ratio for ArcGIS (surface_area.exe). Proceedings of the ESRI International User Conference, San Diego, 13-17 July 2009.http://www. jennessent. com/arc gis/surface_ area.htm

44. Gower, J.C. (1971) A General Coefficient of Similarity and Some of Its Properties. Biometrics, 27, 857-871.

CHAPTER 9

Petroleum Products in Soil Mediated Oxidative Stress in Cowpea (Vigna unguiculata) and Maize (Zea mays) Seedlings

Fidelis Ifeakachuku Achuba

Department of Biochemistry, Delta State University, Abraka, Nigeria

ABSTRACT

The effects of petroleum products (kerosene, diesel, engine oil and petrol) treatment of soil at various sublethal concentrations (0.0%, 0.1%, 0.25%, 0.5%, 1.0%, 1.5% and 2.0%) on oxidative stress markers (lipid peroxidation, superoxide dismutase activity, catalase activity and xanthine oxidase) were studied in cowpea and maize seedlings. The results indicated that the petroleum products caused a significant increase in lipid peroxidation and a significant decrease in the activities of the antioxidant enzymes: Superoxide dismutase, catalase and xanthine oxidase activities. Kerosene had a greater effect on these indicators of oxidative stress than did the other petroleum products. The effects on lipid peroxidation and antioxidant enzymes were more pronounced in cowpea seedlings than in maize seedlings.

KEYWORDS

Cowpea Seedlings, Maize Seedlings, Catalase Activity, Superoxide Dismutase Activity, Xanthine Oxidase Activity

1. INTRODUCTION

In photosynthetic plants, two forms of activated oxygen are formed from superoxide anion. These are hydrogen peroxide and hydroxyl radicals. Activated oxygen is often formed as a component of metabolism to enable "complex" chemical reactions such as the oxidation of xenobiotics or the polymerisation of lignin [1] . It is also formed by the dysfunctioning of enzymes or electron transport systems as a result of perturbations in the metabolism caused by chemical or environmental stress [1] . Activated oxygen species are produced in various sites in plants. These include the chloroplast, peroxisomes and mitochondria [2] - [4] . Various Fe-S proteins and NADH dehydrogenase have also been implicated as possible sites of superoxide and hydrogen peroxide formation [1] . The activated oxygen species generated either during normal metabolic activity or during metabolism of xenobiotics, when produced more than the body can accommodate leads to lipid peroxidation.

This process of lipid peroxidation precedes oxidative damage in plants and animals. However, living organisms are endowed with antioxidant defence systems. The defense mechanisms against oxidative damage include enzymes such as superoxide dismutase, catalase as well as non enzymes such as ascorbic acid. Alterations in the level of these antioxidants represent a measure of oxidative stress. In addition, the activity of xanthine oxidase is also an example of defense mechanism as well as a measure of oxidative stress.

Crude petroleum has been reported to alter oxidative stress indices [5] - [8] . These oxidative stress markers include lipid peroxidation and changes in the activities of anti-oxidant enzymes such as superoxide dismutase, catalase as well as xanthine oxidase The aim of this study was to investigate the effects of petroleum products: Kerosene, diesel, engine oil and petrol contaminated soil on some oxidative stress markers in cowpea and maize seedlings.

2. MATERIALS AND METHODS

2.1. Refined Petroleum Products and Planting Materials

The refined petroleum products of known specific gravities (kerosene = 0.81; diesel = 0.85; engine oil = 0.87; petrol = 0.75) were obtained from Warri Refining and Petrochemical Company, Warri, Nigeria. Improved varieties of maize (Zea mays) were obtained as single batch from Delta Agricultural Development Project (DTADP) Ibusa Delta State, Nigeria. Improved varieties of Vigna unguiculata (L.) Walp were obtained from International Institute of Tropical Agriculture IITA, Ibadan, Nigeria. The soil (sand 84%, silt 5.0%, clay 0.4% and organic matter 0.6%, pH 6.1) was obtained from a fallow land in Delta State University, Abraka. The nutrient content of the soil used is shown in Table 1. The experiment was carried out in a laboratory condition of temperature 28°C and 12 hr day/night regime.

2.2. Soil Treatment and Planting of Seeds

One thousand six hundred grams of soil was added to each small size planting bags (1178.3 cm^3, 15 cm deep) and divided into six groups of five replicates. Groups 1 to 5 contained 0.1%, 0.25%, 0.5%, 1.0% and 2.0% (v/w) respectively of each of the petroleum products while group six served as control (0.0%). To the first bag, 1.6 ml of kerosene, corresponding to 0.1%, was added. The petroleum product treated soil sample was mixed vigo- rously with hand to obtain homogeneity of the mixture. The procedure was repeated for 0.25%, 0.5%, 1.0%, 1.5% and 2.0%. This same procedure was applied to diesel, engine oil and petrol. Each treatment including control was replicated five times. The treatments were watered every day in order to keep the soil moist. The design of the experiment was completely randomized design (CRD).

Table 1. Physicochemical properties of test soil.

Parameters	Value
pH	6.09
Total organic carbon, %	2.90
Phosphorus, mg/kg	<0.01
Nitrogen, mg/kg	8.47
Nitrate, mg/kg	9.86
Cation exchange capacity, meq/100g	0.74
Sodium, mg/kg	9.06
Potassium, mg/kg	6.72
Calcium, mg/kg	2.98
Magnesium, mg/kg	0.31

Damaged seeds were determined by floatation. All seeds that floated on water were discarded and others that remained at the bottom of water were deemed potentially plantable. Three seeds were planted in each test bag to an approximate depth of 2 cm immediately after pollution and kept under partial shade. During the experiment 80 cm^3 of water was supplied to the set up as at when needed to keep the soil moist. Germination [which is indicated by the appearance of epicotyls (for cowpea) and hypocotyls (for maize) above the soil level] records was taken at 4 days interval up to 12 days. Seeds, which failed to sprout after 12 days were regarded as not germinable. At the end of each experimental period, the seedlings were carefully removed from the bags by destroying the bags while the bulk soil containing the seedling was placed under slow running tap water to wash off the soil particles.

2.3. Preparation of Extracts for the Determination of Oxidative Stress Markers in the Leaves of Cowpea and Maize Seedlings

The leaves (0.5 g) of four day old cowpea seedlings was measured and homogenized in pre-hilled mortar with pestle in the presence of 0.05 M phosphate buffer pH 7.5 and few drops of butylated hydroxyl toluene (BHT), filtered with double layered cheese cloth and then centrifuge at 5000 g for 10 min. The supernatant obtained was finally used for the determination of oxidative stress markers. The same procedure was adopted in the preparation of extract for eight and twelve-day-old cowpea seedlings respectively. This same procedure was followed in the preparation of leave extract of four, eight and twelve day's old maize seedlings.

2.4. Determination of Lipid Peroxidation Markers in the Leaves of Cowpea and Maize Seedlings

This assay is based on the reaction of malondialdehyde (MDA) with thiobabituric acid (TBA); forming a MDA-TBA_2 adduct that absorbs strongly at 532 nm. Acetic acid (1.0 ml) was placed in a test tube and to the test tube 1.0 ml of 10% TBA was added followed by 0.1 ml of the supernatant. The test tube was cov- ered and immersed in boiling water for 15 min. the mixture was cool, thereafter centrifuged at 5000 g for 10 min. The spectrophotometer was zeroed and absorbance of test sample was read at 532 nm against the reagent blank.

2.5. Determination of Superoxide Dismutase (SOD) Activity in the Leaves of Cowpea and Maize Seedlings

SOD inhibits the auto-oxidation of epinephrine to adrenochrome [9] . To 2 ml of the homogenate, 2.5 ml of 0.05 M phosphate buffer, pH 7.4 was added. The reaction was initiated by the addition of 0.5 ml of freshly prepared 0.3 nm epinephrine to the buffer-supernatant mixture. This was mixed by inversion. The reference cuvette contained 2.5 ml of the buffer

0.5 ml of epinephrine and 2 ml of deionized water. The increase in absorbance at 480 nm was monitored every second for 150 seconds. One unit of superoxide dismutase activity is defined as the amount of enzyme required for 50% inhibition of the oxidation of epinephrine to adrenochrome at 480 nm per minute [10] . The enzyme activity was assayed with an Sp 1800 UV/VIS Spectrophptometer.

2.6. Determination of the Activity of MnSOD and Cu/Zn SOD in the Leaves of Cowpea and Maize Seedlings

Manganese dependent superoxide dismutase (MnSOD) was analysed in the presence of 1 mM NaCN, to suppress Cu/Zn SOD activity. The cytosolic copper/zinc superoxide dismutase (Cu/Zn SOD) activity was determined, as the difference between total SOD and cyanide sensitive enzyme activity [11] . The enzyme activity was assayed with a spectrometer S22.

The assay mixture contained 1.0 ml of the homogenate, 10 ml of 0.05 M phosphate buffer pH 7.4 and 0.05 ml of 1 mM NaCN. The reaction was initiated by adding 0.5 ml of 0.3% freshly prepared epinephrine. This was mixed by inversion. The absorbance was monitored at 480 nm for every 30 seconds for 150 se- conds.

2.7. Determination of Catalase Activity in the Leaves of Cowpea and Maize Seedlings

Catalase breaks down hydrogen peroxide to give oxygen that oxidizes potassium dichromate. The oxidation of chromate gives a chromophore that absorbs maximally at 610 nm. The enzyme extract (0.5 ml) was added to the reaction mixture containing 1 ml of 0.05 M phosphate buffer (pH 7.5), 0.5 ml of 0.2 M H_2O_2, 0.4 ml H_2O and incubated for different time period t_1, t_2 and t_3 for 1 minute, 2 minutes and 3 minutes respectively. The reaction was terminated after each time interval by the addition of 2 ml of acid reagent (dichromate/acetic acid mixture) which was prepared by mixing 5% potassium dichromate with glacial acetic acid (1:3 by volume). To the control, the enzyme was added after the addition of acid reagent. All the tubes were heated for 10 minutes in boiling water and the absorbance was read at 610 nm with an Sp 1800 UV/VIS Spectrophptometer. Catalase activity was expressed in terms of moles of H_2O_2 consumed/min [12] .

2.8. Determination of Xanthine Oxidase Activities in the Leaves of Cowpea and Maize Seedlings

Xanthine oxidase catalyses the conversion of methylene blue to the reduced clourless forms. Enzyme activity is proportional to the reciprocal of time taken for methylene blue to change to colourless. Two test tubes labeled control and test were placed in a test tube rack, one milliter of 0.05% neutral formaldelyde was pipetted into each test tubes. The 0.02% methylene blue solution was added in the test tube labeled test and followed by the addition of 1 ml of the supernatant to the respective test tube. 1 ml of distilled water was added to the control test tube and 2 drops of liquid paraffin was also added in both test tubes to prevent atmospheric oxidation.

2.9. Statistical Analysis

The results were expressed as mean ± SEM. All results were compared with respect to the control. Comparisons between the test and control were made by using Analysis of Variance (ANOVA), Least Significant Difference (LSD) was used to conduct Post Hoc test for the significant difference. Differences at $p < 0.05$ were considered as significant.

3. RESULTS

Lipid peroxidation in the leaves of cowpea and maize seedlings grown in kerosene, diesel, and engine oil and petrol treated soil after four days of germination are shown in Figure 1. Lipid peroxidation in the leaves of cowpea seedling grown in kerosene treated soil was found to increase significantly ($p < 0.05$) from 0.25% through 2%. In maize seedlings grown in kerosene treated soil , there was a significant ($p < 0.05$) increase in lipid peroxidation in the leaves at 1.5% and 2% concentrations relative to control. Comparing lipid peroxidation in the leaves of cowpea and maize seedlings grown in kerosene treated soil; lipid peroxidation in the leaves was significantly higher in cowpea seedlings than in maize seedlings at 1.5% and 2% concentrations. Lipid peroxidation in the leaves of cowpea seedlings grown in diesel treated soil was found to increase significantly ($p < 0.05$) from 0.25% through 2% concentration compared with control. In maize seedlings grown in diesel treated soil, there was a significant ($p < 0.05$) increase in lipid peroxidation in the leaves from 0.25% to 2% concentration when compared with control. Lipid peroxidation was significantly ($p < 0.05$) lower in the leaves of cowpea seedlings than in the leaves of maize seedlings at 0.1% and slightly higher in the leaves of cowpea seedling than in the leaves of maize seedling at all other levels of concentrations tested. Also, lipid peroxidation in the leaves of cowpea seedlings grown in engine oil treated soil were found to increase significantly ($p < 0.05$) from 0.25% to 2% concentration when compared with control. Lipid peroxidation in the leaves of maize seedling grown in engine oil treated soil was also found to increase significantly from 0.25% to 2%. In comparison, lipid peroxidation was lower in the leaves of cowpea seedlings than in the leaves of maize seedlings. In all the concentrations tested, petrol treated soil resulted in significant ($p < 0.05$) increase of lipid peroxidation from 0.25% to 2% in the leaves of cowpea seedling when compared with control. Lipid peroxidation in the leaves of maize seedlings grown in petrol treated soil also exhibited significant increase from concentration level of 0.5% to 2% when compared with control. When compared, lipid peroxidation in the leaves of cowpea and maize seedlings grown in petrol treated soil was not significantly different. Lipid peroxidation in the leaves of cowpea and maize seedlings grown in kerosene, diesel, and engine oil and petrol treated soil after eight days of germination are shown inFigure 1. Lipid peroxidation in the leaves of cowpea seedlings grown in kerosene treated soil after eight days was found to increase significantly from 0.25% through 2% when compared with control. In maize seedlings grown in kerosene treated soil, there were significant ($p < 0.05$) increases in lipid peroxidation in the leaves at 1.5% and 2% concentrations compare to control. Comparing lipid peroxidation in the leaves of cowpea and maize seedlings grown in kerosene treated soil; lipid peroxidation was significantly ($p < 0.05$) higher in the leaves of cowpea seedling than in the leaves of maize seedlings from 0.25% through 2% concentrations. Lipid peroxidation in the leaves of cowpea seedlings grown in diesel treated soil were found to be significantly ($p < 0.05$) higher from 0.25% to 2% when compared with control. In maize seedlings grown in diesel treated, lipid peroxidation in the leaves showed a significant ($p < 0.05$) increase from 0.5% through 2% concentrations. Lipid peroxidation were significantly ($p < 0.05$) higher in the leaves of cowpea seedling at 1% and 1.5% concentration than in the leaves of maize seedling. Also, lipid peroxidation in the leaves of cowpea seedling grown in engine oil treated soil were found to increase significantly ($p < 0.01$) from 0.25% to 2% engine oil contaminated soil relative to control. Lipid peroxidation in the leaves of maize seedlings grown in engine oil treated soil showed a significant ($p < 0.05$) increase from 1% to 2% concentrations relative to control. In comparison, lipid peroxidation was significantly lower in the leaves of cowpea seedling than in the leaves of maize seedlings. Lipid peroxidation in the leaves of cowpea seedling grown in petrol treated soil was found to increase significantly ($p < 0.05$) from 0.25% to 2% concentrations compared with control. Lipid peroxidation in the leaves of maize seedlings grown in petrol treated soil showed a significant ($p < 0.05$) increase from 0.25% to 2% compared to control. When lipid peroxidation in the leaves of cowpea seedlings and maize seedlings grown in petrol treated soil are compared, lipid peroxidation level was found to be significantly ($p < 0.05$) higher at 0.25% in the leaves of cowpea than in the leaves of maize seedlings. Lipid peroxidation in the leaves of cowpea and maize seedlings grown in kerosene, diesel, and engine oil and petrol treated soil after twelve days of germination are shown in Figure 2. Lipid peroxidation in the leaves of cowpea seedlings grown in kerosene treated soil was found to increase significantly at 0.25% through 2%. In maize seedlings grown in kerosene treated soil, there was a significant increase in leaves lipid peroxidation at 1.0% to 2.0% concentrations respectively of kerosene in soil. Comparing lipid peroxidation in the leaves of cowpea and maize seedlings grown in kerosene treated soil; lipid peroxidation was significantly higher in the leaves of cowpea seedlings than in the leaves of maize seedlings at 0.25% and 2% concentrations. Similarly, lipid peroxidation in the leaves of cowpea seedlings grown in diesel treated soil was found to increase significantly at 0.25% to 2% concentrations compared to control. In maize seedlings grown in diesel treated soil, there was a significant increase in lipid peroxidation in the leaves at 0.25% to 2% concentrations compared to control. Lipid peroxidation was significantly higher from 0.25% to 2% concentrations in the leaves of cowpea seedlings compared to that in the leaves of maize seedling. Also, lipid peroxidation in the leaves of cowpea and maize seedlings grown in engine oil treated soil were found to increase significantly from 0.25% to 2% relative to their respective control values. Maize seedlings grown in engine oil treated soil showed a significant increase in lipid peroxidation in the leaves from 0.25% to 2% concentrations relative to the control. In comparison, lipid peroxidation was significantly lower in the leaves of cowpea seedlings grown in engine oil contaminated soil than in the leaves of maize from 0.25% to 2% concentrations. Lipid peroxidation in the leaves of cowpea and maize seedlings grown in petrol treated soil were found to increase significantly from 0.25% to 2% concentrations. When lipid peroxidation in the leaves of cowpea and maize seedlings are compared in petrol treated soil, lipid peroxidation was found to be significantly higher at 0.25%, 1.0%, and 2.0% concentrations in cowpea seedlings over maize seedlings.

The activities of SOD in the leaves of cowpea and maize after four days of germination in kerosene, diesel, and engine oil and petrol treated soil are presented in Figure 3. Kerosene treated soil resulted in significant decrease ($p < 0.05$) in SOD activity in the leaves of cowpea seedlings compared with control. Maize seedling grown in kerosene treated soil resulted in significant ($p < 0.05$) decrease in SOD activity in the leaves in all the concentrations tested except at 1.5% where a slight decrease was observed. When the activities of SOD in the leaves of cowpea and maize seedlings grown in kerosene treated soil are compared, the enzyme was found to be significantly higher at 0.5% and 1% concentration in the leaves of cowpea seedlings than in the leaves maize seedlings. SOD activities in the leaves of cowpea seedlings grown in diesel treated soil showed a significant ($p < 0.05$) decrease across all the concentrations tested except at 1.5% compared with control. In maize seedlings grown in diesel treated soil, SOD activities in the leaves decreased significantly

($p < 0.05$) across all the concentrations tested relative to the control. When the activities of SOD in the leaves of cowpea and maize seedlings grown in diesel treated soil are compared, it was found to be slightly higher in cowpea seedling than maize seedlings. The activities of SOD in the leaves of cowpea seedling exposed to engine oil treated soil caused a significant ($p < 0.05$) increase in the activity of the enzyme above 1% levels of concentration compared with control. It only caused a slight decrease at 0.25% concentration. In maize seedlings grown in diesel treated soil SOD activity significantly ($p < 0.05$) increased at 1.5% concentration and also showed a slight decrease at 0.1% and 0.25% concentration when compared with control. In all the concentrations, the activities of the SOD in the leaves were significantly higher in cowpea seedlings when compared to maize seedlings. Petrol treated soil had a slight increase in the activity of SOD in the leaves of cowpea seedlings at 1% concentration but showed slight reduction at 0.25% and 0.5% concentrations when compared with control. In maize seedlings however, only slight decrease in the activity of leaves SOD was observed at 0.25% concentration. No reasonable change was shown at all other levels of concentration. In all the concentrations tested, the activities of the enzyme were found to be significantly ($p < 0.05$) higher in the leaves of cowpea seedlings compared to maize seedlings. After eight days, kerosene treated soil resulted in significant decrease in SOD activity in the leaves of cowpea seedlings compared with control. Similarly, kerosene treated soil caused a significant decrease in SOD activity in the leaves of maize seedlings. When the activities of SOD in the leaves of cowpea and maize seedlings grown in kerosene treated soil are compared, the enzyme was found to be significantly higher in the leaves of cowpea seedlings at kerosene concentration of 1% to 2% than in the leaves of maize seedlings. The activities of SOD in the leaves of cowpea seedlings grown in diesel treated soil showed a significant ($p < 0.05$) decrease at 0.1% to 0.5% concentrations when compared with control. Slight decrease of leaves SOD activity was observed at other levels of concentration except 1%. Kerosene treated soil resulted in significant ($p < 0.05$) decrease of SOD activity in the leaves of maize seedlings in all the concentrations tested compared with control. When the activities of SOD in the leaves of cowpea and maize seedlings grown in diesel treated soil are compared; it was found to be significantly ($p < 0.05$) higher in the leaves of cowpea seedlings than in the leaves of maize seedlings.

SOD activity in the leaves of cowpea seedling grown in engine oil treated soil showed a significant ($p < 0.01$) decrease up to 0.25%. However, further increase in engine oil concentration led to significant ($p < 0.05$) increase of leaves SOD activity. However, in maize seedlings grown in engine oil treated soil, there was a significant ($p < 0.05$) increase in the activity of leaves SOD at 1.5% concentration while a slight decrease of leaves SOD activity at 0.5% concentration was observed. In all the concentrations, the activities of leaves SOD were significantly higher in cowpea seedlings at concentration levels above 0.25% when compared to the leaves SOD of maize seedlings. Petrol treated soil had a slight increase in the activities of total SOD in the leaves of cowpea seedlings at 0.1% and 1% levels of soil contamination. However, a significant decrease of leaves SOD activity in cowpea seedling at 0.25% concentration was observed. In maize seedlings, significant ($p < 0.05$) increases of leaves total SOD activity was observed at 0.1% and 1% concentrations; while a significant ($p < 0.05$) decrease was recorded at 0.5%. When compared, leaves total SOD activity in cowpea seedlings was significantly ($p < 0.05$) higher than in maize seedlings. The activities of superoxide dismutase (SOD) in the leaves of cowpea and maize seedlings after twelve days of germination in kerosine, diesel, engine oil and petrol treated soil are shown in Figure 3. In all the concentrations tested, kerosene treated soil resulted in significant decrease ($p < 0.05$) of SOD activity in the leaves of cowpea compared with control. However, in maize seedling, kerosene treated soil only resulted in significant decrease ($p < 0.05$) up to 0.5% concentration; thereafter further increase in kerosene concentration did not result in significant decrease in SOD activity. When the activities of SOD in the leaves of cowpea and maize seedlings grown in kerosene treated soil are compared, the enzyme was found to be significantly lower in the leaves of cowpea seedlings at kerosene concentration of 1.0% than in the leaves of maize seedlings. Similarly, in all the concentrations tested, diesel treated soil resulted in significant decrease ($p < 0.05$) of SOD in the leaves of cowpea seedlings compared with control. However, in diesel treated soil, there was significant decrease in the activity of SOD in the leaves of maize seedlings at 0.25%. Moreover, there was a significant increase in the activity of SOD in the leaves of maize seedlings at 0.1, 1.0% 1.5% and 2% diesel concentrations. When the activities of total SOD in the leaves of maize seedlings are compared with those of cowpea seedlings grown in diesel treated soil, the enzyme was found to be significantly ($p < 0.05$) higher in the leaves of maize seedling at diesel concentrations above 0.5%. Unlike kerosene treated soil, the changes in the activities of the enzyme in diesel treated soil are not concentration dependent.

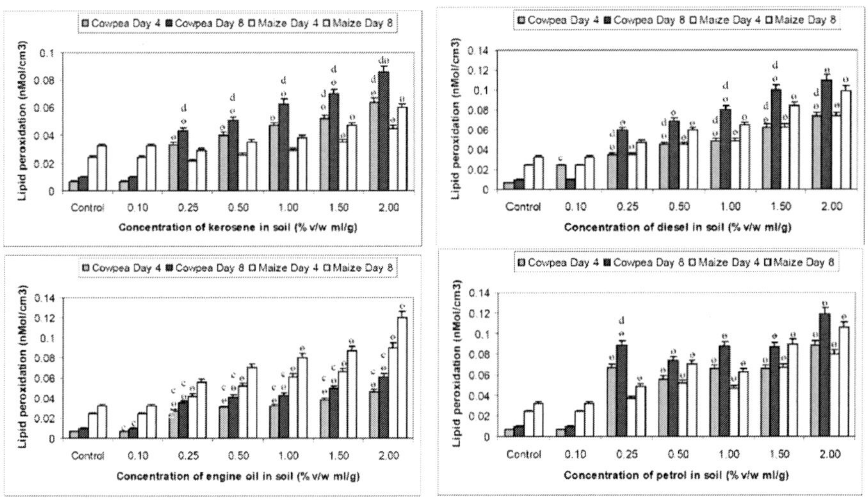

Figure 1. Effect of concentration of petroleum products on lipid peroxidation in the leaves of cowpea and maize after four and eight days of germination. [*]Significantly lower as compared to control; [+]Significantly lower as compared to engine oil; [++]Significantly lower as compared to kerosene; [e]Significantly lower relative to control; [a]Significantly lower relative to other petroleum products; [b]Significantly higher relative to other petroleum products; [c]Significantly lower in cowpea relative to maize seedlings; [d]Significantly higher in cowpea relative to maize seedlings.

The activities of superoxide dismutase (SOD) in the leaves of cowpea seedling grown in engine oil treated soil significantly ($p < 0.05$) decreased up to 0.5%. At higher concentrations of engine oil above 1.0% there was a significant increase in the activity of the enzyme in the leaves of cowpea seedlings. However, in maize seedling grown in engine oil treated soil, there was a significant ($p < 0.05$) increase in the activity of leaves SOD in all the concentrations except at 0.25% engine oil in soil. In all the concentrations, the activities of the enzymes were significantly ($p < 0.05$) lower in the leaves of cowpea seedlings when compared to maize seedling. Petrol treated soil had a significant ($p < 0.05$) reduction in the activities of (SOD) in the leaves of cowpea seedlings grown in all the concentrations tested relative to the control. In maize seedlings grown in petrol treated soil, there was a significant ($p < 0.05$) increase in the activity of leaves SOD at 0.1%, 0.5%, 1.0%, 1.5% and 2.0% petrol in soil. However, there was no change in the activity of the enzyme in maize seedlings grown in 0.25% petrol contaminated soil. In all the concentrations tested the activity of leaves SOD was found to be significantly lower in the leaves of cowpea seedlings compared to that in the leaves of maize seedlings.

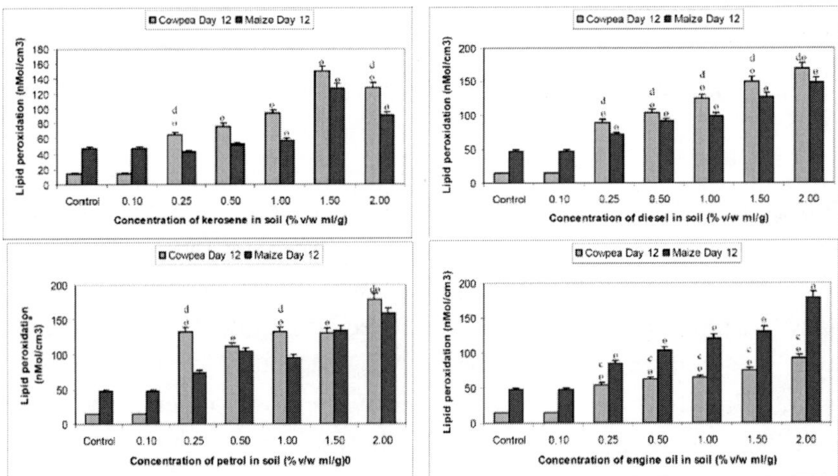

Figure 2. Effect of concentration of petroleum products on lipid peroxidation in the leaves of cowpea and maize after twelve days of germination. [*]Significantly lower as compared to control; [+]Significantly lower as compared to engine oil; [++]Significantly lower as compared to kerosene; [e]Significantly higher relative to control; [a]Significantly lower relative to other petroleum products; [b]Significantly higher relative to other petroleum products; [c]Significantly lower in cowpea relative to maize seedlings; [d]Significantly higher in cowpea relative to maize seedlings.

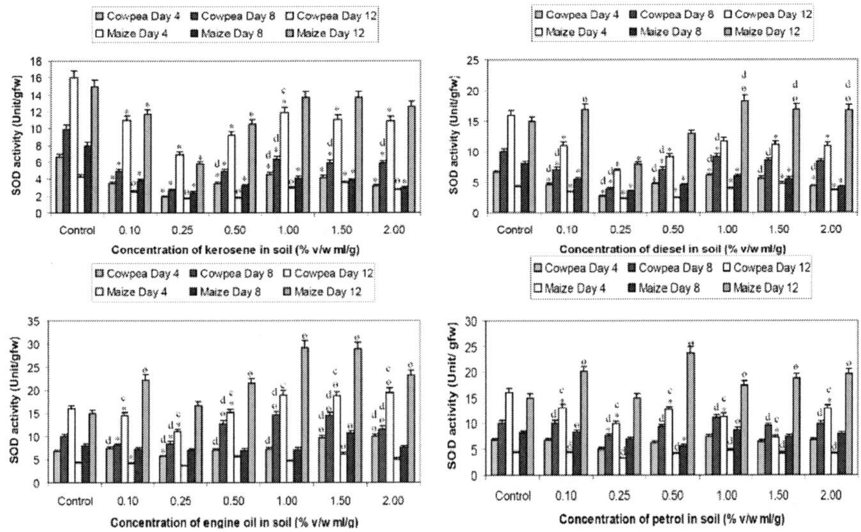

Figure 3. Effect of concentration of petroleum products on total SOD activities in leaves of cowpea and maize after four, eight and twelve days of germination. [*]Significantly lower as compared to control; [+]Significantly lower as compared to engine oil; [++]Significantly lower as compared to kerosene; [e]Significantly higher relative to control; [a]Significantly lower relative to other petroleum products; [b]Significantly higher relative to other petroleum products;[c]Significantly lower in cowpea relative to maize seedlings; [d]Significantly higher in cowpea relative to maize seedlings.

The activities of Cu/Zn SOD in the leaves of cowpea and maize seedlings after four days of germination in kerosene, diesel, and engine oil and petrol treated soil are shown in Figure 4. In all the concentrations tested, kerosene treated soil showed a significant decrease of Cu/Zn SOD activities in the leaves of cowpea and maize seedlings when compared with their respective control values. When the activities of Cu/Zn SOD in the leaves of cowpea seedlings grown in Kerosene treated soil were compared to that in maize seedlings, it was found to be significantly higher in the leaves of cowpea seedlings at 1% to 2.0% than in the leaves of maize seedlings. In cowpea seedlings grown in diesel treated soil, the activities of Cu/ZnSOD in the leaves showed a significant ($p < 0.05$) decrease in all the concentrations tested compared with control. However, in maize seedlings a significant ($p < 0.05$) decrease in Cu/ZnSOD activity in the leaves at 0.1%, 0.25%, 0.5% and

2% concentrations was observed. However, leaves Cu/ZnSOD exhibited a slight decrease at 1% and 1.5% concentrations compared with control. In contrast, the activities of Cu/ZnSOD in the leaves of cowpea seedlings grown in diesel treated soil was significantly ($p < 0.05$) lower than the activity in the leaves of maize seedlings in all the concentrations above 0.5%. At the concentrations tested the activities of Cu/Zn SOD in the leaves of cowpea seedlings grown in engine oil treated soil slightly decreases at 0.25% and 0.5% concentrations. Maize seedlings grown in engine oil treated soil exhibited a significant ($p < 0.01$) increase of Cu/Zn SOD activities in the leaves at 1.5% concentration. A slight decrease however, was observed at 0.25% and 1% concentrations. By way of contrast, the activities of Cu/Zn SOD in the leaves of cowpea seedlings grown in engine oil treated soil appeared to be significantly ($p < 0.05$) higher than in the leaves of maize seedlings. In all the concentrations tested, cowpea seedlings grown in petrol treated soil showed a slight decrease of Cu/Zn SOD activities in the leaves at 0.1%, 0.25% and 2% concentrations. However, a significant ($p < 0.01$) decrease in the activity of leaves Cu/Zn SOD was observed at other levels of concentrations tested. In maize seedlings grown in petrol treated soil the activities of leaves Cu/Zn SOD decreased slightly at all concentrations tested except at petrol concentration of 1.5% where it showed a significant ($p < 0.05$) decrease in the activities of the enzyme. When the activities of Cu/Zn SOD in the leaves of cowpea seedlings are compared with the activities of the enzyme in maize seedlings, it was observed to be significantly ($p < 0.05$) lower in the leaves of cowpea seedling. The activities of Cu/Zn SOD activity in the leaves of cowpea and maize seedlings after eight days of germination in kerosene, diesel, engine oil and petrol treated soil are shown in Figure 4. In all the concentration tested, kerosene treated soil showed a significant ($p < 0.05$) decrease of Cu/Zn SOD activities in the leaves of both cowpea and maize seedlings compared with control. When the activities of Cu/Zn SOD in the leaves of cowpea and maize seedlings grown in kerosene treated soil are compared, it was found to be significantly ($p < 0.05$) higher in cowpea seedlings at levels of soil contamination above 0.25%. Diesel treated soil, resulted in significant ($p < 0.05$) decrease of Cu/Zn SOD activity in the leaves of cowpea seedlings in all the concentrations tested except 1.5% compared with control. Similarly, maize seedlings grown in diesel treated soil showed a significant ($p < 0.05$) decrease in the activities of Cu/Zn SOD in the leaves at all levels of concentration tested except 1% diesel concentration. When compared, diesel treated soil resulted in a significantly ($p < 0.05$) higher activity of Cu/Zn SOD in the leaves of cowpea seedlings than in maize seedlings. At the various concentrations; the activity of Cu/Zn

SOD in the leaves of cowpea seedlings grown in engine oil treated soil significantly ($p < 0.05$) increase at 1.5% but slight decreases at concentration levels below 1.5% compared with control were observed. The activity of Cu/Zn SOD in the leaves of maize seedlings grown in engine oil treated soil significantly ($p < 0.05$) increase at 1% concentration. By way of contrast, the activities of Cu/Zn SOD were significantly higher in the leaves of cowpea seedlings than in maize seedling. In all the concentrations tested, cowpea seedlings grown in petrol treated soil had Cu/Zn SOD activities in the leaves slightly lower at 0.1%, and 1% concentrations, but significantly ($p < 0.05$) decreased at other levels of concentration when compared with control. In maize seedlings grown in petrol treated soil, a significant ($p < 0.05$) increase in the activities of Cu/Zn SOD in the leaves at 0.1%, 0.5% and 1% concentration was observed but minor increases were recorded at 0.25% and 1.5%. When compared, there was no significant difference in the activities of Cu/Zn SOD in the leaves of both cowpea and maize seedlings grown in petrol treated soil. The activity of copper/zinc superoxide dismutase (Cu/ZnSOD) in the leaves of cowpea and maize seedlings after twelve days of germination in kerosene, diesel, engine oil and petrol treated soil are shown in Figure 4. In cowpea seedlings grown in kerosene treated soil a significant decrease of Cu/ZnSOD activity in the leaves was observed at all concentrations tested. Similarly, in maize seedlings grown in kerosene contaminated soil a significant decrease of Cu/ZnSOD activity in the leaves at 0.1%, 1.0%, 1.5% and 2.0% was observed. When the activities of Cu/Zn SOD in the leaves of cowpea seedlings grown in kerosene treated soil were compared to those of maize seedlings, non significant lower values in cowpea across the various levels of concentrations was observed. In cowpea seedlings grown in diesel treated soil, the activities of Cu/Zn SOD in the leaves were found to be significantly lower at all concentrations tested compared to the control. At the various concentrations, the activities of Cu/Zn SOD in the leaves of maize seedling grown in diesel treated soil exhibited an increase at 0.1%, and 1.5%; a decrease in the activities of Cu/Zn SOD in the leaves of maize seedlings grown in diesel contaminated soil was shown at other concentrations except at 0.25% concentration. The activities of Cu/Zn SOD in the leaves of cowpea seedling grown in diesel treated soil exhibited decreases at various concentrations compared to the activity in the leaves of maize seedlings. The activities of Cu/ZnSOD in the leaves of cowpea seedling grown in engine oil treated soil significantly ($p < 0.05$) decreased at 0.1%, to 2%, but increased at 1.5%, compared to control. However, in maize seedling grown in diesel treated soil, very high increases of Cu/ZnSOD activities in the leaves at 0.1%, 0.5% to 2% concentrations were observed. The activities of Cu/Zn SOD in the leaves of cowpea seedlings grown in engine oil treated soil, were significantly ($p < 0.05$) lower across the various concentrations than in maize seedlings. Cowpea seedlings grown in petrol treated soil had Cu/Zn SOD activities in the leaves relatively lower at 0.1%, and 0.5% concentrations respectively. However, petrol treated soil resulted in a decrease in Cu/Zn SOD in the leaves activities at 0.25%, 1%, 1.5% and 2.0% concentrations. Maize seedlings grown in petrol treated soil only resulted in significant increase in the activities of Cu/Zn SOD in the leaves at all concentrations except at 1% and 1.5% relative to control. When compared, the activities of the enzyme in the leaves of cowpea seedlings significantly ($p < 0.05$) than in the leaves of maize seedlings.

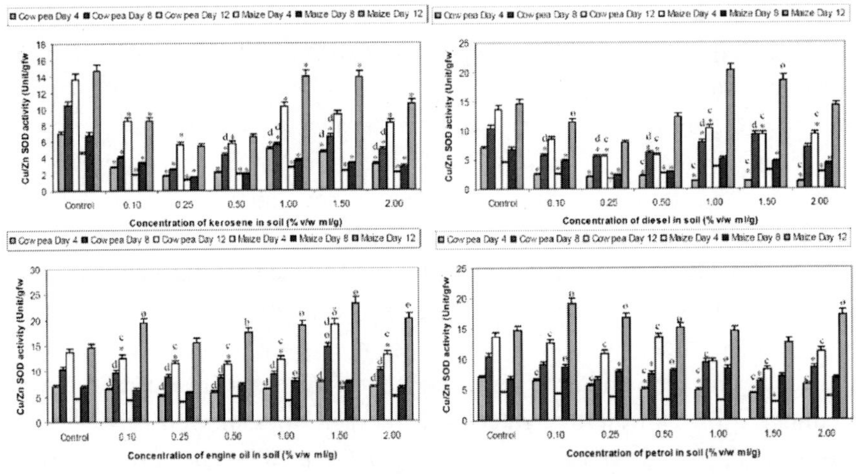

Figure 4. Effect of concentration of petroleum products on Cu/Zn SOD activities in leaves of cowpea and maize after four, eight and twelve days of germination *Significantly lower as compared to control; +Significantly lower as compared to engine oil; ++Significantly lower as compared to kerosene; °Significantly higher relative to control; [a]Significantly lower relative to other petroleum products; [b]Significantly higher relative to other petroleum products; [c]Significantly lower in cowpea relative to maize seedlings; [d]Significantly higher in cowpea relative to maize seedlings.

The activities of MnSOD in the leaves of cowpea and maize seedlings after four days of germination in kerosene, diesel, and engine oil and petrol treated soil are shown in Figure 5. Kerosene treated soil resulted in significant ($p < 0.05$) decrease of MnSOD activities in the leaves of cowpea seedlings except at kerosene concentration of 1%. Similarly, kerosene treated soil gave rise to a significant decrease of MnSOD activities in the leaves of maize seedlings at all the levels of concentrations tested except 1.5% where a slight increase was observed. There was a higher increase in MnSOD activities in the leaves of cowpea seedlings than in the leaves of maize seedlings when compared. The activities of MnSOD in the leaves of cowpea seedlings grown in diesel treated soil significantly ($p < 0.05$) decreased at 0.1% 0.25% and 0.5% concentrations compared to control. However, a slight decrease was recorded at concentration levels above 1%. The activities of MnSOD in the leaves of maize seedlings exposed to diesel showed a slight decrease at all levels of concentrations tested except at 0.25% and 0.5% where significant ($p < 0.05$) reduction was observed. At 1.5% concentration MnSOD activity in the leaves of maize seedling showed a slight increase when compared with control. Nonetheless, there was a higher increase in MnSOD activities in the leaves of cowpea seedlings than in the leaves of maize seedlings. Engine oil treated soil resulted in significant ($p < 0.05$) increase in MnSOD activities in the leaves of cowpea seedlings at concentration levels above 1%. It was also found that MnSOD activities in the leaves activities exhibited a significant ($p < 0.05$) decrease at 0.1% and 0.25% concentrations. Also, MnSOD activities in the leaves of maize seedlings grown in engine oil treated soil significantly ($p < 0.05$) increased at 1.5% and 2% concentrations. However, a slight decrease was recorded at 0.25% concentration. In addition, the activities of MnSOD were significantly ($p < 0.05$) higher in the leaves of cowpea seedlings than in the leaves of maize seedlings when compared. Petrol treated soil resulted in slight decrease of MnSOD activities in the leaves of cowpea in all the concentrations except at 0.1% and 0.5% concentrations where significant ($p < 0.05$) decreases were observed. However, it showed a significant decrease at 0.25%, 1% and 1.5% concentrations. Also, there was an insignificant increase of MnSOD activities in the leaves at 2% level of petrol concentration. In maize however, petrol treated soil only gave rise to slight increase in MnSOD activities in the leaves in all the levels of concentrations tested except at

1.5% where a minor decrease was observed. When compared, MnSOD activities were slightly higher in the leaves of cowpea seedlings than in the leaves of maize seedlings. The activities of MnSOD in the leaves of cowpea and maize seedlings after eight days of germination in kerosene, diesel, engine oil and petrol treated soils are shown in Figure 5. Kerosene treated soil resulted in significant ($p < 0.05$) decrease of MnSOD activities in the leaves of cowpea seedlings grown in kerosene treated soil compared with control. In maize seedlings, however, MnSOD activities in the leaves decreased significantly ($p < 0.01$) at all the levels of concentrations tested except 1% where a slight reduction was observed. In comparison, the activities of MnSOD were significantly ($p < 0.05$) higher in the leaves of cowpea seedling than in the leaves of maize seedlings. At the concentrations tested, the activities of Mn SOD in the leaves of cowpea seedlings grown in diesel treated soil resulted in a significant ($p < 0.05$) decrease in the activity of the enzyme relative to control value. In maize seedlings exposed to diesel treated soil. MnSOD activity in the leaves decreased significantly ($p < 0.05$) in all the concentrations tested. However, at 1% concentration a slight decrease in MnSOD activity in the leaves was recorded. When the activities of MnSOD in the leaves of cowpea and maize seedlings grown in diesel treated soil were compared, it was found to be significantly ($p < 0.05$) higher across the concentrations tested in cowpea seedlings than in maize seedlings. The activities of MnSOD in the leaves of cowpea seedlings grown in engine oil treated soil significantly ($p < 0.05$) increased at 1.5% and 2% concentrations compared to control value. However, MnSOD activities in the leaves of cowpea seedlings grown at 0.1%, 0.5% and 1% concentration showed a slight increase compared to control. Nevertheless, a significant decrease was observed at 0.25%. In maize seedlings grown in engine oil treated soil, MnSOD activities in the leaves activities showed a significant ($p < 0.05$) increase at 1.5% and 2%. However, there was significant ($p < 0.05$) decrease at 0.1% and 0.25%. Moreover, a slight increase of MnSOD activities in the leaves was also observed at 0.5% and 1% concentrations when compared to the control. When MnSOD activities in the leaves of cowpea and maize seedlings grown in engine oil treated soil were compared, it was observed to be significantly ($p < 0.05$) higher at all the levels of concentrations tested in cowpea than in maize seedlings. Petrol treated soil gave rise to a slight decrease in MnSOD activities in the leaves at concentration levels above

0.25% in cowpea seedlings when compared with control. However, in maize seedling grown in petrol treated soil, a significant decrease of MnSOD activities in the leaves was exhibited in all the concentrations tested. When compared, MnSOD activities were significantly higher in the leaves of cowpea seedlings than in maize seedlings. The activities of MnSOD in the leaves of cowpea and maize seedlings after twelve days of germination in kerosene, diesel, engine oil and petrol treated soils are shown in Figure 5. Kerosene treated soil gave rise to a significant ($p < 0.05$) decrease in Mn SOD activities in the leaves of cowpea seedling grown in all the concentrations except 1% kerosene contaminated soil. The results also show that MnSOD activities in the leaves of cowpea seedlings grown in kerosine contaminated significantly ($p < 0.05$) decreased at 0.25% and 0.5% level of contamination. In maize seedlings grown in kerosene contaminated soil, the MnSOD activities significantly ($p < 0.05$) decreased at concentrations corresponding to those of cowpea seedlings and vice-versa. Nonetheless, MnSOD activities in the leaves of cowpea seedlings were lower than in the leaves of maize seedlings. Diesel treated soil resulted in a significant ($p < 0.05$) decrease in MnSOD activities in the leaves of cowpea seedlings at 0.1%, 1%, 1.5% and 2%. It was also found that MnSOD activities exhibited a significant decrease at 0.25% and 0.5%. Also, MnSOD activities in the leaves of maize seedlings grown in diesel treated soil significantly increased at 1% and 1.5% but decreased at 0.25% and 0.5%. When compared MnSOD activities were lower in the leaves of cowpea than in the leaves of maize seedlings, MnSOD activities in the leaves of cowpea seedling in engine oil treated soil showed a significant increase from 0.5% - 2% concentrations. On the other hand, the activities of Mn SOD in the leaves of maize seedling grown in engine oil treated soil showed a sharp increase at 0.5% - 2% concentrations. The activities of Mn SOD in the leaves of cowpea seedling grown in petrol treated soil resulted in a slight increase at 0.1%, 0.5% and 2% concentrations but decreases were recorded at 0.25%, 1% and 1.5%. The activities MnSOD in the leaves of maize seedling showed a sharp increase at concentrations levels of 0.1%, 0.5% and 2%. Conversely, the result shows a decrease in Mn SOD activities in the leaves at 0.25%, 1% and 1.5% concentrations. However, MnSOD activities were lower in the leaves of cowpea seedlings than in the leaves of maize seedling.

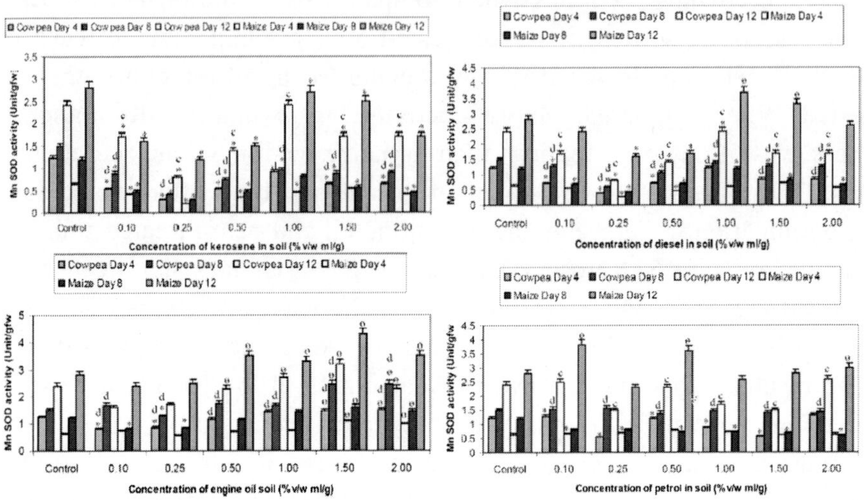

Figure 5. Effect of concentration of petroleum products on MnSOD activities in leaves of cowpea and maize after four, eight and twelve days of germination. *Significantly lower as compared to control; +Significantly lower as compared to engine oil; ++Significantly lower as compared to kerosene; eSignificantly higher relative to control; aSignificantly lower relative to other petroleum products; bSignificantly higher relative to other petroleum products; cSignificantly lower in cowpea relative to maize seedlings; dSignificantly higher in cowpea relative to maize seedlings.

The activities of catalase in leaves of cowpea and maize seedling in kerosene, diesel, and engine oil and petrol treated soil after four days of germination are shown in Figure 6. At the various concentrations tested, kerosene treated soil brought about a successive decrease of catalase activity in the leaves of cowpea seedlings as well as in the leaves of maize seedlings compared with control. In comparison, however, there was no significant difference in the decrease across the various concentrations tested between the activities of catalase in the leaves of cowpea and maize seedlings. Catalase activities tested in the leaves of cowpea seedlings grown in diesel treated showed a decrease in activities at all levels of soil contamination compared to control. Similarly, catalase activity in the leaves of maize seedlings grown in diesel treated indicated a progressive decrease across all levels of concentrations tested. However, catalase activity appeared to be slightly higher in the leaves of cowpea seedlings than in the leaves of maize. In all the concentrations tested, engine oil treated soil resulted in successive decrease of catalase activity in the leaves of cowpea seedlings as well as in the leaves of maize seedlings relative to control. Nonetheless, the decrease of catalase activity was slightly higher in the leaves of cowpea seedlings compared to that of maize seedlings. Petrol treated soil resulted in slight decrease of catalase activity in the leaves of cowpea seedlings at all levels of concentration compared with control. Similarly, in maize seedlings, catalase activity in the leaves showed a minor decrease in all the concentration tested compared to control. There was no significant difference in the activities of catalase in the leaves of cowpea and maize seedlings.

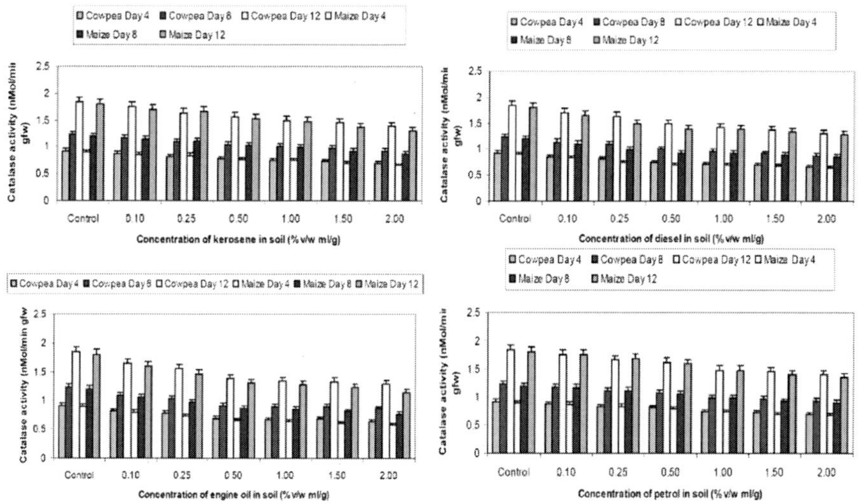

Figure 6. Effect of concentration of petroleum products on catalase activities in leaves of cowpea and maize after four, eight and twelve days of germination. *Significantly lower as compared to control; +Significantly lower as compared to engine oil; ++Significantly lower as compared to kerosene; eSignificantly higher relative to control; aSignificantly lower relative to other petroleum products; bSignificantly higher relative to other petroleum products; cSignificantly lower in cowpea relative to maize seedlings; dSignificantly higher in cowpea relative to maize seedlings.

The activities of catalase in the leaves of cowpea and maize seedling after eight days of germination in kerosene, diesel, engine oil and petrol treated soil are shown in Figure 6. At the various concentrations tested, kerosene treated soil brought about a successive decrease of catalase activity in the leaves of cowpea seedlings as well as in maize seedlings compared with control. In comparison, however, the decreases across the various concentrations tested were more pronounced in the leaves of cowpea seedlings than in the leaves of maize seedlings. Similarly, catalase activity in the leaves of both cowpea seedlings and maize seedlings grown in diesel treated soil decreased throughout the various levels of concentrations tested relative to control. No significant differences existed in the activities of catalase in the leaves of cowpea seedlings and maize seedlings grown in diesel treated soil. In all the concentrations tested, engine oil treated soil resulted in a slight decrease of catalase activity in the leaves of cowpea seedlings compared with control. In maize seedlings, catalase activity in the leaves insignificantly reduced across all the concentrations tested relative to control. By contrast, the decrease of catalase activity was slightly more in the leaves of maize seedling than cowpea seedlings. In all the concentrations tested, petrol treated soil gave rise to a decrease of catalase activity in the leaves of cowpea seedling compared with control. Similarly, petrol treated soil brought about decrease of catalase activity in the leaves of maize seedlings. When compared, the decrease in the activities of catalase was more in the leaves of maize seedlings than in the leaves of cowpea seedlings

The activities of catalase in the leaves of cowpea and maize seedlings after twelve days of germination in kerosene, diesel, engine oil and petrol treated soil are shown in Figure 6. At the various concentrations tested, kerosene treated soil brought about a successive decrease of catalase activity in the leaves of cowpea seedlings as well as in the leaves of maize seedlings compared with control. There were no distinct differences between the activities of the enzyme across the various concentrations in the leaves of cowpea seedlings as well as in the leaves of maize seedlings. Catalase activities in the leaves of cowpea seedlings grown in diesel treated soil showed a decrease in activities in all the concentrations tested. Similarly, catalase activities in all the concentrations tested in the leaves of maize seedling indicated a progressive decrease. However, catalase activities decreased less in the leaves of cowpea seedlings relative to that in the leaves of maize seedlings. Also in all the concentrations tested, engine oil treated soil resulted in significant decrease of catalase activity in the leaves of cowpea seedlings compared with control. Similarly, catalase activities in the leaves of maize seedling decreased successively as concentration of engine oil increase. However, the decrease in catalase activity in the leaves of cowpea seedlings was slightly lower compared to that of maize seedlings. In all the concentrations tested, petrol treated soil gave rise to a significant ($p < 0.05$) decrease in catalase activity in the leaves of cowpea seedling. Similarly, catalase activities decreased in the leaves of maize seedling in all the concentrations tested relative to control. However, the reduction in the catalase activity at 0.1%, 1% and 2% concentrations was less in the leaves of cowpea seedlings than in the leaves of maize seedlings. On the contrary, the reduction of catalase activities in the leaves of cowpea seedlings grown in petrol treated soil at 0.25% was more than that of maize seedlings.

The activities of xanthine oxidase in the leaves of cowpea and maize seedlings after four days of germination in kerosene, diesel, engine oil and petrol treated soil are shown in Figure 7. In kerosene treated soil, of the various concentrations tested, it was found that xanthine oxidase activities in the leaves of cowpea and maize seedlings decreased significantly ($p < 0.05$) compared with control. Comparing xanthine oxidase activities in the leaves of cowpea and maize seedlings, there was no significant difference between the activities of the enzyme at each level of concentration. In diesel treated soil, xanthine oxidase activity decreased significantly ($p < 0.05$) across all the concentrations tested in the leaves of both cowpea and maize seedlings. By comparison, the activity of xanthine oxidase in the leaves of cowpea seedlings appeared to be higher at 0.1% - 0.5% and lower at 2% concentrations in relation to the xanthine oxidase activity in the leaves of maize seedling. In all the concentrations tested, engine oil treated soil resulted in significant decrease in xanthine oxidase activity in the leaves of cowpea seedlings. In same vain, xanthine oxidase activity in the leaves of maize seedlings decreased significantly across all concentrations tested compared with control. In comparison, xanthine oxidase activity was slightly higher in the leaves of cowpea seedlings than in the leaves of maize seedlings, but a significant difference in the enzyme activity was not observed. Petrol treated soil caused a significant ($p < 0.05$) decrease in the activities of xanthine oxidase activity in the leaves of both cowpea and maize seedlings compared with control. In comparison, xanthine oxidase activity was slightly higher in the leaves of cowpea leaves but not significantly different from the activity in the leaves of maize seedlings.

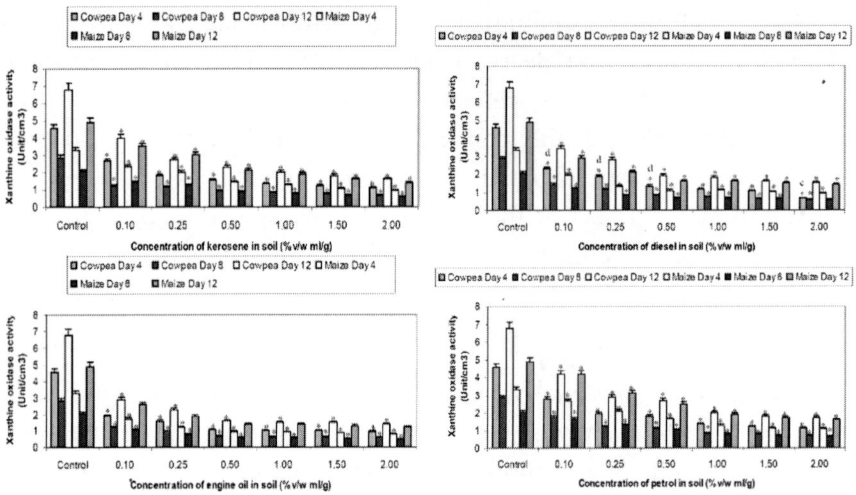

Figure 7. Effect of concentration of petroleum products on xanthine oxidase activities in the leaves of cowpea and maize after four, eight and twelve days of germination. *Significantly lower as compared to control; +Significantly lower as compared to engine oil; ++Significantly lower as compared to kerosene; eSignificantly higher relative to control; aSignificantly lower relative to other petroleum products; bSignificantly higher relative to other petroleum products; cSignificantly lower in cowpea relative to maize seedlings; dSignificantly higher in cowpea relative to maize seedlings.

The activities of xanthine oxidase in the leaves of cowpea and maize seedlings after eight days of germination in kerosene, diesel, engine oil and petrol treated soil are shown in Figure 7. In kerosene treated soil, of the various concentrations tested, it was found that xanthine oxidase activities in the leaves of maize seedlings decreased significantly ($p < 0.05$) compared with control. Similarly, xanthine oxidase activity in the leaves of cowpea seedlings grown in kerosene treated soil significantly

($p < 0.05$) decrease in all the concentrations tested. Comparing xanthine oxidase activities in the leaves of cowpea and maize seedlings, it was found to be slightly lower in cowpea seedlings at all concentration except at 0.1% and 0.25%, but no significant difference existed in xanthine oxidase activity between cowpea and maize seedlings. In diesel treated soil, xanthine oxidase activities decreased significantly ($p < 0.05$) across all the concentration tested in the leaves of both cowpea and maize seedlings. By comparison the activities of xanthine oxidase increased slightly in the leaves of cowpea seedlings than in the leaves of maize seedlings across all the concentrations tested. In all the concentrations tested, engine oil treated soil resulted in significant ($p < 0.05$) decrease in xanthine oxidase activity in the leaves of cowpea seedlings. Similarly, xanthine oxidase activity decreased significantly in the leaves of maize seedlings grown in engine oil treated soil compared with control. In comparison, the activities of xanthine oxidase in the leaves of cowpea seedlings are higher than in maize seedlings. However, there is no significant difference between the xanthine oxidase activity in the leaves of cowpea and maize seedlings. Petrol treated soil gave rise to significant ($p < 0.05$) decrease in xanthine oxidase activity in the leaves of cowpea seedlings compared with control. Similarly, the leaves of maize seedlings grown in petrol treated soil witnessed a significant ($p < 0.05$) decrease in xanthine oxidase activities compared with control. When compared, the activities of xanthine oxidase activities in the leaves of cowpea seedlings were slightly higher than in the leaves of maize seedlings at 0.1%, 0.5% and1.5%, but slightly lower at 0.25%. The activities of xanthine oxidase in the leaves of cowpea and maize seedlings after twelve days of germination in kerosene, diesel, engine oil and petrol treated soil are shown in Figure 7. In kerosene treated soil, of the various concentrations tested, it was found that xanthine oxidase activities in the leaves of cowpea and maize seedlings decrease significantly ($p < 0.05$) compared with control. Comparing xanthine oxidase activities in the leaves of cowpea and maize seedlings, there was no significant difference between the activities of the enzyme at each level of concentration. In diesel treated soil, xanthine oxidase activity decreased significantly ($p < 0.05$) across all the concentration tested in the leaves of both cowpea and maize seedlings. By comparison, the activity of xanthine oxidase in the leaves of cowpea seedling grown in diesel treated soil was higher than in the leaves of maize seedling. But at each concentration there was no significant difference between the activity of the enzyme in the leaves of both cowpea and maize seedlings. In all the concentrations tested, engine oil treated soil resulted in a significantly ($p < 0.05$) decrease in xanthine oxidase in the leaves of cowpea seedling. Similarly, it was found that the activity of xanthine oxidase in the leaves of maize seedling was significantly ($p < 0.05$) inhibited across all concentrations relative to control. In comparison, xanthine oxidase activity was slightly higher in the leaves of cowpea seedling but not statistically different from the activity in the leaves of maize seedling. Petrol treated soil caused a significant ($p < 0.05$) decrease of xanthine oxidase activity in the leaves of both cowpea and maize seedlings. In comparison, xanthine oxidase activity was slightly higher in the leaves of cowpea seedling except at 0.25% but not statistically different from the activity in the leaves of maize seedlings.

4. DISCUSSION

The formation of lipid peroxidation products in plants exposed to adverse environmental conditions is an indication of free radical formation in tissue and it may be used as index of lipid peroxidation in biological system [13] [14] . Previous report indicated that metal exposure results in the generation of reactive oxygen species in plant [15] . The present results show that an increasing level of refined petroleum products in soil increased the level of lipid peroxidation in the leaves of both cowpea and maize seedlings exposed to petroleum products treated soil for four and eight days after germination (Figure 1) and twelve days (Figure 2). An increase in the level of lipid peroxidation has been reported for plants exposed to metal ion [14] [16] - [18] . Elevated levels of reactive oxygen species initiate lipid peroxidation [19] [20] that culminate in oxidative stress [21] . An increase in the level of lipid peroxidation is the evidence most frequently cited in support of the involvement of oxidative stress in tissues [22] -[24] .

This study shows that refined petroleum products (Kerosene, diesel, and petrol and engine oil) could induce lipid peroxidation in exposed plants (Figure 1 and Figure 2). This is because refined petroleum toxicity could result in the production of reactive oxygen species (ROS); which in turn can cause membrane damage. The relationship between lipid peroxidation and ROS has been reported [25] [26] . The process of lipid peroxidation was more pronounced in cowpea seedlings compared to maize seedlings (Figure 1 and Figure 2), which also lends further credence to the report that monocotyledonus seeds are less affected by toxicants than dicotyledonous seeds.

The production of reactive oxygen species during stress has been reported to results from pathways such as photorespiration, from the photosynthetic apparatus and mitochondria respiration. Environmental stresses have been shown to trigger the active production of reactive oxygen species by NADPH oxidase [27] - [30] . The enhanced production of reactive oxygen species during stress can pose threat to cells but plant have inbuilt mechanism to counteract the reactive oxygen species. The scavenging mechanisms of plant include superoxide dismutase (SOD) and catalase [31] - [33] . The present investigation indicated that the exposure of cowpea and maize seedlings to refined petroleum products, kerosene, diesel engine oil and petrol affected superoxide dismutase activity after four, eight days and twelve days of germination (Figure 3). After twelve days of exposure to kerosene, diesel and petrol treated soil the activity of total SOD decreased relative to control but there was increase in the activity of the enzyme in the leaves of cowpea and maize seedlings grown in engine oil treated soil compared to the other petroleum products (Figure 3). Decrease in the activity of the enzyme portends reduction in the capacity of the exposed plant to handle reactive oxygen species. This is because increase in the activity of the antioxidant enzyme may have a role in imparting tolerance against any type of environmental stress [33] . Moreover, the petroleum products caused a reduction in Cu/Zn-dependent superoxide dismutase after four, eight days and twelve days of germination (Figure 4) as well as Mn-containing superoxide dismutase (MnSOD) after four, eight days and twelve days of growth in contaminated soil (Figure 4). These enzymes are involved in the general defense system against natural or chemically induced production of reactive oxygen species [34] . It has been suggested that a decrease in the activity in these detoxification mechanisms can generate severe cell damage due to imbalance in the production of toxic oxygen radicals. Conversely, increased activity could contribute to cell protection from chemical toxicants [35] - [38] .

Smirnoff [39] established that an increase in the capacity of antioxidant defenses in response to an increased level of reactive oxygen represents an indirect measure of oxidative stress.

The results indicated that kerosene affected the seedlings more than the other refined petroleum products. Similarly, the activity of the antioxidant enzymes in the leaves of maize seedlings appears to be less affected by the petroleum products compared to cowpea seedlings. This is exhibited by the effects of petroleum products on CuZnSOD (Figure 4) and MnSOD (Figure 5) in cowpea and maize seedlings. This is in agreement with previous reports in which maize seedling has been shown to be resistant to crude oil contamination [40] while beans possess the ability to uptake crude oil [41] . However, this observation is in contrast to the report of Baek et al. [42] in which Zea mays was found to be more sensitive to oil exposure than red beans. Previous reports of interspecies differences in sensitivity to petroleum hydrocarbon, which may be related to differences in systemic uptake of hydrocarbon and cell wall structure, had been documented [43] [44] .

Catalase is another important enzyme that protect living system against oxidative stress, being able to scavenge hydrogen peroxide, which is the major product produced by superoxide dismutase [45] . The inhibition of catalase activity as concentration of petroleum product increases after four, eight days and twelve days of germination (Figure 6) corroborates the effect of petroleum hydrocarbon on superoxide dismutase activity in the leaves of both cowpea and maize seedlings. The similarity of catalase to superoxide dismutase is not suprising because the two enzymes have been reported to act in tandem [5] [46] . However, there is no marked difference in the activity of catalase in the leaves of both cowpea and maize seedlings at each of the concentration tested (Figure 6).

A wealth of information is available confirming that xanthine oxidase is involved in the metabolism of heterocyclic and polycyclic aromatic hydrocarbons [47] [48] as well as the involvement of the enzyme in the conversion of hypoxanthine via xanthine into uric acid. The role of uric acid as an antioxidant has been documented [49] [50] . The significant ($p < 0.05$) decreases in the activity of the enzyme as the concentration of petroleum products increased after four, eight and twelve days of germination (Figure 7) portends that petroleum mediated decrease in xanthine oxidase activity in cowpea and maize seedlings could contribute to oxidative stress by its decreased ability to metabolize aromatic hydrocarbon as well as its reduced ability to generate uric acid, a potent antioxidant. This effect is more pronounced in the leaves of cowpea seedlings compared to maize seedlings (Figure 7). The decrease in the activities of superoxide dismutase, catalase and xanthine oxidase as a result of increased formation of reactive oxygen species was earlier reported [51] - [54] . Therefore, it is pertinent to note that refined petroleum products could impose oxidative stress in exposed plant.

5. CONCLUSION

This study has indicated that exposure of plant to refined petroleum products in soil could impose oxidative stress on plant. The effect was more on cowpea relative to maize seedlings. Moreover, kerosene affected the exposed plant more than the other petroleum products

REFERENCES

1. Mittler, R. (2002) Oxidative Stress, Antioxidants and Stress Tolerance. Trends in Plant Science, 7, 405-410.

2. Elstner, E.F. (1991) Mechanisms of Oxygen Activation in Different Compartments of Plant Cells. In: Pell, E.J. and Steffen, K.L., Eds., Active Oxygen/Oxidative Stress and Plant Metabolism, American Society of Plant Physiologists, Rockville, 13-25.

3. Voet, D. and Voet, J.G. (1995) Biochemistry. 2nd Edition, John Wiley and Sons, New York, 816-820.

4. Rich, P.R. and Bonner Jr., W.D. (1978) The Site of Superoxide Anion Generation in Higher Plant Mitochondria. Archive of Biochemistry and Biophysics, 188, 206-213.

5. Achuba, F.I. and Osakwe, S.A. (2003) Petroleum Induced Free Radical Toxicity in African Catfish (Clarias garieponus). Fish Physiology and Biochemistry, 29, 97-103.

6. Anozie, O.I. and Onwurah, I.N. (2001) Toxic Effect of Bonny Light Crude Oil in Rat after Ingestion of Contaminated Diet. Nigerian Journal of Biochemistry and Molecular Biology (Proceeding Supplement), 16, 1035-1085.

7. Downs , C.A. , Shigenska, G., Fauth, J.E., Robinson, C.E. and Huang, A. (2002) Cellular Physiological Assessment of Bivalves after Chronic Exposure to Spolled Exon Valdes Crude Oil Using a Novel Molecular Diagnostic Biotechnology. Environmental Science and Technology, 36, 2987-2993.

8. Achuba, F.I., Peretiomo-Clarke, B.O. and Ebokaiwe, P. (2005) Pollution Induced Oxidative Stress in African Catfish (Clarias heterobranchus). European Journal of Scientific Research, 8, 52-73.

9. Aksnes, A. and Njaa, R.L. (1981) Catalase, Glutathione Peroxidase and Superoxide Dismutase in Different Fish Species. Comparative Biochemistry and Physiology, 69B, 893-896.

10. Misra, H.P. and Fridovich, I. (1972) The Role of Superoxide in the Auto-Oxidation of Epinephrine and a Simple Assay for Superoxide Dismutase. Biochemical Journal, 247, 3170-3175.

11. Crapo, J.D., McCord, J.M. and Fridovich, I. (1978) Preparation and Assay of Superoxide Dismutases. Methods in Enzymology, 53, 382-393.

12. Rani, P., Meena Unni, K. and Karthikeyan, J. (2004) Evaluation of Antioxidant Properties of Berries. Indian Journal of Clinical Biochemistry, 19, 103-110.

13. Heath, R.L. and Packer, L. (1968) Photoperoxidation in Isolated Chloroplasts. I. Kinetics and Stoichiometry of Farry Acid Peroxidation. Archives Biochemistry and Biophysics, 125, 189-198.

14. Srivastava, M., Lena, Q.M., Singh, N. and Singh, S. (2005) Antioxidant Responses of Hyper-Accumulator and Sensitive Fern Species to Arsenic. Experimental Botany, 56, 1335-1342.

15. Hartley-Whitaker, J., Ainsworth, G. and Mehary, A.A. (2001) Copper and Arsenate-Induced Oxidative Stress in Hocus lanatus L. Clones with Differential Sensitivity. Plant Cell and Environment, 24, 713-722.

16. Somashekaraiah, B.V., Padmaja, K. and Prasad, A.R.K. (1992) Phytotoxicity of Cadmium Ions on Germinating Seedlings of Mung Bean (Phaseolus vulgaris): Involvement of Lipid Peroxides in Chlorophyll Degradation. Physiologia Plantarum, 85, 85-89.

17. Gallego, S.M., Benavides, M.P. and Tomaro, M.L. (1996) Effect of Heavy Metal Ion Excess on Sunflower Leaves: Evidence for Involvement of Oxidative Stress. Plant Science, 121, 151-159.

18. Lozano-Rodriguez, E. Hernandez, C.E., Bonay, P. and Carpena-Ruiz, R.O. (1997) Distribution of Cadmium in Shoots and Root Tissues of Maize and Pea Plants: Physiological Disturbances. Journal of Experimental Botany, 48, 123-128.

19. Frei, B. (1994) Reactive Oxygen Species and Antioxidant Vitamins: Mechanism of Action. American Journal of Medicine, 97, S5-S13.

20. Val, A.L. and Almeida-Val, V.F. (1999) Effects of Crude Oil on Respiratory Aspect of Some Fish Species of the Amazon. In: Val., A.L. and Almerda-Val, V.M.F., Eds., Biology of Tropical Fish, INPA, Manaus, 227-291.

21. Yu, Q. and Rengel, Z. (1999) Micronutrient Deficiency Influences Plant Growth and Activities of Superoxide Dismutase in Narrow Leafed Lupins. Annals of Botany, 8, 175-182.

22. Halliwell, B. (1989) Oxidants and the Central Nervous System: Some Fundamental Questions. Acta Neurologica Scandinavica, 126, 23-33.

23. Halliwell, B. and Gutteridge, J.M.C. (1990) The Antioxidant of Human Extra Cellular Fluids. Archives of Biochemistry and Biophysics, 280, 1-8.

24. Liu, J. and Mori, A. (1994) Involvement of Reactive Oxygen Species in Emotional Stress: A Hypothesis Based on the Immobilization Stress Induced Oxidative Damage and Antioxidant Defense Changes

in Rat Brain and the Effect of Antioxidant Treatment with Reduced Glutathione. International Journal of Stress Management, 1, 249-263.

25. Achuba, F.I. (2010) Spent Engine Oil Mediated Oxidative Stress in Cowpea (Vigna unguiculata) Seedlings. Electronic Journal of Environment, Food and Agricultural Chemistry, 9, 910-917.

26. Gutteridge, M.C. (1995) Lipid Peroxidation and Antioxidants as Biomarker of Tissue Damage. Clinical Chemistry, 41, 1819-1828.

27. Hammond-Kosack, K.E. and Jones, J.D.G. (1996) Resistance Gene-Dependent Plant Defense Responses. The Plant Cell Online, 8, 1773-1791.

28. Orozo-Cardenase, M. and Ryan, C.A. (1999) Hydrogen Peroxide Is Generated Systematically in Plant Leaves by Wounding and Systemin via the Octadecanoid Pathway. Proceedings of the National Academy of Sciences of the United States of America, 96, 6553-6557.

29. Cazale, A.C., Droillard, M.J., Wilson , C., Heberle-Bors, E., Barbier-Brygoo, H. and Lauriere, C. (1999) MAP Kinase Activation by Hypoosmotic Stress of Tobacco Cell Suspensions: Towards the Oxidative Burst Response? Plant Journal, 19, 297-307.

30. Pei, Z.M., Murata, Y., Benning, G., Thomine, S., Klüsener, B., Allen, G.J., Grill, E. and Schroeder, J.I. (2000) Calcium Channels Activated by Hydrogen Peroxide Mediate Abscisic Acid Signalling in Guard Cells. Nature, 406, 731-734.

31. Asada, K. and Takahashi, M. (1987) Production and Scavenging of Active Oxygen in Chloroplasts. In: Kyle, D.J., Osmond, C.B. and Arntzen, C.J., Eds., Photoinhibition, Elsevier, Amsterdam, 227-287.

32. Bowler, C., Van Montague, M. and Inze, D. (1994) Superoxide Dismutase in Plants. Critical Review of Plant Science, 13, 199-218.

33. Jayakumar, K., Jaleel, A.C. and Viayarengan, P. (2007) Changes in Growth, Biochemical Constituents and Antioxidant Potentials in Radish (Raphasus sativus L.) under Cobalt Stress. Turkish Journal of Biology, 31, 127-136.

34. Fridovich, I. (1986) Biological Effects of Superoxide Radical. Archive of Biochemistry and Biophysics, 247, 1-11.

35. Sevenian, A. and Hochistein, P. (1985) Mechanism and Consequences of Lipid Peroxidation in Biological Systems. Annual Review of Nutrition, 5, 365-390.

36. Stern, A. (1985) Red Cell Oxidative Damage. In: Sies, H., Ed., Oxidative Stress, Academic Press, New York, 331-349.

37. Parke, V.O. (1987) Role of Enzymes in Protection against Lipid Peroxidation. Regulatory Toxicology and Pharmacology, 7, 222-235.

38. Saltman, P. (1989) Oxidative Stress: A Radical View. Seminar in Hematology, 26, 249-256.

39. Smirnoff, N. (1993) The Role of Active Oxygen in the Response of Plants to Water Deficit and Desiccation. New Phytologist, 125, 27-58.

40. Ayotarmuno, J.M. and Kogbara, R.B. (2007) Determining the Tolerance Level of Zea mays (Maize) to Crude Oil Polluted Agricultural Soil. African Journal of Biotechnology, 6, 1332-1337.

41. Aboaba, O.A., Aboaba, O.O., Nwachuku, N.C., Chukwu, E.E. and Nwachukwu, S.C.U. (2007) Evaluation of Bioremediation of Agricultural Soils Polluted with Crude Oil by Planting Beans Seeds, Phaseolus vulgaris. Nature and Science, 5, 53-60.

42. Baek, K., Kom, H., Oh, H., Young, B., Kim, J. and Lea, I. (2004) Effect of Crude Oil Components and Bioremediation on Plant Growth. Journal of Environmental Science and Health, A39, 2465-2477.

43. Albert, P.H. (1995) Petroleum and Individual Polycyclic Aromatic Hydrocarbons. In: Hoffman, D.J., Rattner, B.A., Burton Jr., G.A. and Cairons Jr., J., Eds., Handbook of Ecototoxicology, CRC Press, Boca Raton, 330-355.

44. Ogbo, E.M. (2009) Effects of Diesel Fuel Contamination on Seed Germination of Four Crop Plants. Arachis hypogala, Vigna

unguiculata, Sorghum bicolor and Zea mays. African Journal of Biotechnology, 8, 250-253.

45. Asada, K. (1992) Ascorbate Peroxidase: A Hydrogen Peroxide Scavenging Enzyme in Plants. Physiologia Plantarum, 85, 235-241.

46. Iyawe, H.O.T. and Onigbinde, A.O. (2004) Effect of an Anti-Malarial and Micronutrient Supplementation on Respiration Induced Oxidative Stress. Pakistan Journal of Nutrition, 3, 318-321.

47. Panoutsopoulos, G.I. and Beedham, C. (2004) Enzymatic Oxidation of Phthalazine with Guinea Pig Liver Aldehyde Oxidase and Liver Slices: Inhibition by Isovanillin. Acta Biochemistry Polonica, 51, 953-951.

48. Panoutsopolous, G.I., Kauretas, D. and Beedham, C. (2004) Contribution of Aldehyde Oxidase, Xanthine Oxidase and Aldehyde Dehydrogenase on the Oxidation of Aromatic Aldehyde. Chemical Research in Toxicology, 17, 1368-1376.

49. Peden, D.B., Hohman, R., Brown, M.E., Masont, R.T., Berkebilef, C., Falest, H.M. and Kaliner M.A. (1990) Uric Acid Is a Major Antioxidant in Human Nasal Airway Secretions. Proceedings of National Academy of Science, 87, 7638- 7642.

50. Nieto, F.J., Iribarren, C., Gross, M.D., Comstock, G.W. and Cutler, R.G. (2000) Uric Acid and Serum Antioxidant Capacity: A Reaction to Atherosclerosis? Atherosclerosis, 148, 131-139.

51. Trush, M.A. and Kensler, T.W. (1991) An Overview of the Relationship between Oxidative Stress and Chemical Carcinogenesis. Free Radical Biology and Medicine, 10, 201-209.

52. Sies, H. (1991) Oxidative Stress: Introduction. In: Sies, H., Ed., Oxidative Stress: Oxidant and Antioxidants, Academic Press, San Diego, 15-22.

53. Pigeolet, E., Corbisler, P., Houbion, A., Lambert, D., Michiels, C., Raes, M., et al. (1990) Glutathione Peroxidases, Superoxidase and Oxygen Derived Free Radicals. Mechanisms of Ageing and Development, 51, 283-297.

54. Achuba, F.I. and Otuya, E.O. (2006) Protective Influence of Vitamins against Petroleum-Induced Free Radical Toxicity in Rabbit. The Environmentalist, 26, 295-300.

CHAPTER 10

Tile Drainage and Nitrogen Fertilizer Management Influences on Nitrogen Availability, Losses, and Crop Yields

Rakesh Awale[1], Amitava Chatterjee[1], Hans Kandel[2], Joel K. Ransom[2]

[1]Soil Science, North Dakota State University, Fargo, ND, USA

[2]Plant Sciences, North Dakota State University, Fargo, ND, USA

ABSTRACT

Installing tile drainage facilitates early planting and field operations, and tiling has tremendously increased in the Red River Valley (RRV) due to recent wet cycles. This experiment studied tile drainage and N-fertilizer management effects on N availability, N losses, and yields of corn (Zea mays L.) and sugarbeet (Beta vulgaris L.) in a naturally poorly-drained Fargo soil during the 2012-2013 growing seasons. Regardless of drainage, applying 146 kg N ha^{-1} with nitrapyrin resulted similar soil N availability to 180 kg N ha^{-1} without nitrapyrin in sugarbeet in both years. In corn, application of nitrapyrin resulted either higher or similar soil N levels to split-N application. In 2013, application of urea alone increased soil N availability during the early corn growing season under the undrained condition, whereas nitrapyrin delayed N release in the tile-drained soils. Corn and sugarbeet yields averaged 7.4 and 47.0 $Mg \cdot ha^{-1}$ in 2012, and averaged 8.3 and 38.3 $Mg \cdot ha^{-1}$ in 2013, respectively, with no significant differences among N-sources. However, corn yield increased on an average by 27.6% with N-fertilizer application over unfertilized control in 2013. In 2012, sugarbeet root impurity (% sucrose loss to molasses) increased by 13.8% and 17.2% with 146 kg N ha^{-1} plus nitrapyrin and 180 kg N ha^{-1} treatments, respectively, compared to unfertilized control. Besides, higher N rates were usually associated with greater daily soil N_2O emissions, with the maximum flux of 105 g N_2O-N $ha^{-1} \cdot d^{-1}$ recorded under corn. Addition of fertilizer-N increased NH_3 volatilization losses up to 1.9% and 0.5% of the applied-N in corn and sugarbeet, respectively. Tile drainage influenced soil N availability more than crop yield during two years of study. Nitrogen management can have pronounced effects on N availability and losses. A long-term study is needed to investigate the fertilizer-N use efficiency of crops under tile drainage condition.

KEYWORDS

Subsurface Drainage, Nitrogen, Nitrous Oxide, Ammonia Volatilization, Nitrate Leaching

1. INTRODUCTION

About 1.8 million hectares of soils in the RRV of North Dakota and northern Minnesota are naturally poorly drained [1] , which has led to increased adoption of subsurface tile drainage [2] . Tile drainage can provide agronomic benefits through the gravimetric water removal, improved trafficability, and timeliness of field operations [3] . However, shifting water regimes influences below ground N dynamics through changes in soil moisture and temperature [4] .

Soil water content has significant influence on N mineralization rates [5]. Tile drainage can accelerate soil N mineralization by improving soil aeration [6] . Optimum N rates for crop production can vary greatly among years based on N mineralization rates and possible N losses [7] . Fertilizer-N recommendations are usually based on the amount of N available at the beginning of growing season. However, in soils with high organic matter, a substantial amount of soil N mineralization is deemed possible during the crop growing season, which is less considered by the growers [8] . Therefore, monitoring mineral N contents in soils during the growing season can improve our knowledge on N availability to crops.

Denitrification losses of applied N increase with soil water content because of root-zone oxygen reduction through displacement of soil air by water [9] . Installing tile drainage can eliminate the chance of water logging by lowering the water table, and thereby, the denitrification N losses [10] . Apart from soil aeration, the availability of inorganic N substrates is also an important factor that controls the dynamics of soil nitrous oxide (N_2O) emissions [11] . Denitrification loss of N_2O from agricultural soils increases with N-application rates [12] . Large pulses of N_2O productions at times immediately following fertilizer N application and/or heavy rainfall events are not uncommon [9] - [12] .

Tile drainage has potential to reduce denitrification loss of N_2O. However, nitrate (NO_3^-) leaching could be a major concern in tile-drained soils [13] . Annual rainfall, crop yield variations, and soil and nutrient management, in part or combination, govern NO_3^- leaching loss [14] . Greater N losses to subsurface drains are observed in years following a drought due to greater residual N in the soil profile [3] [13] .

Tile drainage can also influence ammonia (NH_3) volatilization by changing soil water and temperature regimes, and/or influencing N substrate availability; however, the extent of control mechanism is still not understood [10] . Ammonia volatilization is a major N loss pathways associated with the use of surface applied NH_4^+ based fertilizers [12] . Soil water regimes have a strong impact on the NH_4^+ transport through profile and subsequently on the availability of NH_4^+ substrate for NH_3 volatilization [15] . Using tile drainage to lower soil water table provides surface soil water with more volume to percolate down the soil profile that can incorporate fertilizer N into the soil, resulting in low

NH_3 volatilization [16] . Conversely, tile drainage may increase NH_3volatilization by maintaining the soil water content at field capacity [17] .

Soil N supply is essential for crop yield and quality [18] . Insufficient N limits crop yield, but excessive amounts of N can deteriorate air and water quality [12] - [15] . Application of fertilizer-N has generally shown to increase crop yields but it also increases N losses via N_2O emissions, NO_3leaching, and NH_3 volatilization, especially when fertilizer-N is applied in excess of crop requirements [12] . Application of commercial nitrification inhibitor compounds such as nitrapyrin (NP) delay NH_4^+ oxidation to NO_3^-, and increase N availability to the crops during the periods of rapid crop N uptake [19] . Application of N fertilizers in split doses could be another approach to co-ordinate maximum crop N uptake [18] . Reducing the availability of inorganic N in soils when crop N requirement is small, N-inhibitors and split N application have the potential to increase N use efficiency and yields of crops, while minimizing potential N losses [8] [12] [14] [20] .

Balancing the amount of N required for optimum crop growth while minimizing potential N losses under tile drainage conditions is a major challenge in the RRV. Simultaneous measurements of crop yields, N availability, and N losses as influenced by interaction of N management and tile drainage have not been intensively studied in the RRV. We hypothesize that crop yields would be increased under tile drainage condition due to increase in the N availability under favorable soil water level. Main objectives of this field experiment were to determine tile drainage and fertilizer-N management practices effect on (i) corn and sugarbeet yields, (ii) changes in the soil N availability (iii) denitrification loss of N in the form of N_2O, (iv) soil solution NO_3^- concentration at 60 cm soil depth, and (v) NH_3 volatilization loss of N in a poorly drained Fargo silty clay soil.

2. MATERIALS AND METHODS

2.1. Description of Experimental Site

Field experiments were located at North Dakota State University research site (46.93°N, 96.85°W) near Fargo, North Dakota, USA. Broadly, the site has Fargo silty clay soil and is classified as Fine, smectitic, frigid Typic Epiaquerts with 0% - 1% slope [21] . Subsurface drainage tiles were installed at the 2.5 ha experimental area in 2008. The area was divided into eight units of 61 m long by 54 m wide, each unitconsisted of seven lateral subsurface drainage tile lines. Tile lines of 10 cm in diameter were installed at a depth of 90 cm with a spacing of 7.6 m, and with a drainage coefficient of 7.5 mm·d^{-1}.

Each unit was controlled via a water table control structure (Agri-Drain Corp, Adair, IA, USA). Four of the units had the control structures open to represent subsurface drainage and the remaining four units had the control structures closed to represent undrained field conditions.

2.2. Field Experiments and Experimental Design

At the experimental site, corn and sugarbeet were grown during the 2012-2013 growing seasons. A randomized complete block design was used with four replicates in a split-plot arrangement with drainage (undrained and tile-drained) as the main plot factors and N fertilizer management as the sub plot factors in each crop for both years. For corn, the N treatments in 2012 were (i) 180 kg N ha^{-1} applied at preplant as urea (Urea180), (ii) 224 kg N ha^{-1} at preplant as urea (Urea224), (iii) 224 kg N ha^{-1} at preplant as urea plus nitrapyrin (Urea224 + NP), (iv) 112 kg N ha^{-1} at preplant as urea plus 112 kg N ha^{-1} at 6 leaf stage as urea ammonium nitrate (UAN) [Split (Urea112 + UAN112)]. The N-rates of 180 kg N ha^{-1} and 224 kg N ha^{-1} were selected on the basis of corn yield goals of 8.4 and 10.4 Mg·ha^{-1}, respectively, within this region [22] . In 2013, the N-rate was lowered to 134 kg N ha^{-1} considering residual soil N from fall soil test results, soybean N credits and corn N requirement for the yield goal of 10.4 Mg·ha^{-1}. The N treatments for corn in 2013 were (i) control (0 N), (ii) 134 kg N ha^{-1} at preplant as urea (Urea224),

(iii) 134 kg N ha^{-1} at preplant as urea plus nitrapyrin (Urea134 + NP), (iv) 67 kg N ha^{-1} at preplant as urea plus 67 kg N ha^{-1} at 6 leaf stage as UAN [Split (Urea67 + UAN67)] in 2013. For sugarbeet, the N management included (i) control (0 N), (ii) 146 kg N ha^{-1} at preplant as urea (Urea146), (iii) 146 kg N ha^{-1} at preplant as urea plus nitrapyrin (Urea146 + NP), and (iv) 180 kg N ha^{-1} at preplant as urea (Urea180) in both years. The N-rate of 146 kg·ha^{-1} is considered as the recommended N-rate for sugarbeet production within this region [22] .

2.3. Field Operations

During 2012 and 2013 growing seasons, corn was planted in the experimental field sections that were under soybean in the previous years. The plots were cultivated to a depth of 7.6 cm once in the fall following the harvest of soybean, and again in the spring before planting corn with a one-pass field cultivator. The preplant urea fertilizers were uniformly broadcasted by hand and incorporated immediately with the field cultivator on April 26 in 2012 and on May 15 in 2013. The nitrification inhibitor nitrapyrin (trade name Instinct, 17.6% by weight active ingredient solution, DOW AgroSciences LLC, IN, USA) was mixed with urea and applied to the soil, based on an area basis at the rate of 2.5 L·ha^{-1} (450 g a.i. ha^{-1}). Corn hybrid PH-8640 RIB (DuPont Pioneer, IA, USA) was planted on April 30 in 2012 and on May 15 in 2013 at the rate of 88000 seeds ha^{-1} with a 1010 John Deere seed planter (John Deere, Moline, IL, USA). The seeds were placed 3.8 cm deep with 15 cm in-row (seed to seed) spacing and 76 cm between-rows (row to row) spacing, such that four corn rows were included in an individual corn plot size of 6.1 m long by 3.0 m wide. For the split N treatments, the UAN was side-dress applied on June 4 in 2012 and on June 18 in 2013. Crops were grown under rainfed condition without any irrigation water inputs. Weeds were controlled using herbicides—applied twice during each corn growing seasons. The first application consisted of amixture of glyphosate (Monsanto, St. Louis, MO, USA] at 1.5 L·ha^{-1} + Status (dicamba and diflufenzopyr, and isoxadifen safener) (BASF, Research Triangle Park, NC, USA) at 13.1 g of a.i. ha^{-1} at V3 (three collar leaf) stage. The second application consisted of glyphosate at 1.5 L·ha^{-1} + Luadis (atrazine tembotrione and isoxadifen safener) (Bayer CropSciene

LP, Research Triangle Park, NC, USA) at 96 ml of a.i. ha^{-1} at V7 (seven collar leaf) stage. All of the corn rows were machine harvested after physiological maturity on Oct. 9, and Oct. 24, respectively, in 2012 and 2013. Corn grain yields were determined at the moisture contents of 15.0%.

In 2012, sugarbeets were grown in the field sections previously under corn whereas in 2013, wheat preceded sugarbeets. The required rates of urea fertilizers were uniformly broadcasted by hand on May 10 and May 29, respectively in 2012 and 2013. Nitrapyrin was mixed with urea and applied to the soil, based on an area basis at the rate of 2.5 $L \cdot ha^{-1}$. The fertilizers were then incorporated using a Triple K field cultivator with rolling basket. On the same day, sugarbeet variety Crystal 985 Roundup Ready (American Crystal Sugar Company, MN, USA) was planted with a John Deere Max Emerge II planter to an individual sub-plot size of 6.1 m long by 3.4 m wide. The seeds were placed 3.2 cm deep with 56 cm row spacing and 7.6 cm in-row spacing. The plots were thinned manually to maintain a plant population of 156500 plants ha^{-1} for the first year only. The beets were grown under rainfed condition without any irrigation water inputs. Glyphosate herbicide was applied on June 22 in 2012, and on July 6 in 2013 at 3.5 $L \cdot ha^{-1}$. Two middle rows from each plot were machine harvested on Sept. 17 in 2012, whereas the beets were harvested manually (3.1 m long each from two middle rows) on Oct. 24 in 2013. The beets were weighed instantly (gross sugarbeet root yield) and subsamples of the sugarbeet roots were sent to American Crystal Sugar Quality Tare Lab, East Grand Forks, Minnesota, USA for yield determinations and quality analyses. From this data, net sugarbeet root yield was calculated after subtracting the external root impurities (tare %) from the gross sugarbeet root yield. The net sugarbeet root yield hereafter is referred to as sugarbeet (root) yield.

2.4. Basic Soil Properties

Before planting sugarbeets, three soil cores—up to a depth of 120 cm with depth intervals of 0 - 30, 30 - 60, and 60 - 120 cm—were collected using a truck mounted probe (3.6 cm internal diameter) and composited per replicate unit in order to determine initial soil inorganic N levels [23]

. Also, separate soil cores (2 cm diameter) were taken from the upper 30 cm soil surface from each individual sugarbeet plots to determine bulk density [24] , organic matter [25] , soil pH and EC [26] , cation exchange capacity [27] , particle size analysis [28] , Olsen-P [29] , and available K [30] . The basic soil physical and chemical properties are presented in Table 1.

2.5. Growing Season Soil Inorganic N Content

After planting, soil samples were collected by hand using a soil probe (2 cm internal diameter) from the upper 30 cm soil profile—with 15 cm increments for both growing season in each crop. In 2012, soil samples were collected at a monthly interval, whereas soil samples were taken at a bi-weekly interval for the first two months and then at a monthly interval until harvest in 2013. Four soil cores were collected from between crop rows and composited for each sub-plot. The samples were transferred to the laboratory in a cooler and stored at −4˚C until analyzed. In the laboratory, soil inorganic N (NH_4^+ and NO_3^-) contents were determined [23] . Field moist soil (6.5 g) was extracted with 25 mL of 2M KCl after shaking the mixture for 30 min in a reciprocal shaker. The soil suspension was then centrifuged for 5 min and filtered through a Whatman no. 2 filter paper. The extracts were then analyzed for inorganic N using an Automated Timberline TL2800 Ammonia Analyzer (Timberline Instruments, CO, USA). Three additional soil cores (2 cm internal diameter) with depth intervals of 0 - 15, and 15 - 30 were also collected using a soil probe per each replicate from each crop at the first soil sampling events in both years to determine bulk density [24] . The average bulk densities from each crop were used to calculate their respective growing season soil inorganic N contents. Soil NH_4^+ and NO_3^- concentrations (mg·kg^{-1}) in the 0 - 15 cm and 15 - 30 cm depth were multiplied by respective (depth-wise) bulk densities to express them into area basis (kg·ha^{-1}). The inorganic N contents at the two depth intervals were summed to obtain NH_4^+ and NO_3^- contents for the upper 30 cm soil profile. Finally, both the NH_4^+ and NO_3^- contents were added together to obtain total inorganic N contents for 0 - 30 cm soil depth.

Table 1. Basic soil characteristics in the surface 0 - 30 cm depth at the experimental site measured in sugarbeet plot during the 2012 and 2013 growing seasons.

Soil properties[†]	2012	2013
Sand ($g \cdot kg^{-1}$)	17 ± 3[‡]	17 ± 5
Silt ($g \cdot kg^{-1}$)	359 ± 47	374 ± 61
Clay ($g \cdot kg^{-1}$)	624 ± 89	609 ± 77
Bulk density ($Mg \cdot m^{-3}$)	1.22 ± 0.11	1.23 ± 0.06
pH	8.24 ± 0.27	8.32 ± 0.38
EC ($dS \cdot m^{-1}$)	1.59 ± 0.56	1.54 ± 0.62
Organic matter ($g \cdot kg^{-1}$)	71.8 ± 1.7	70.7 ± 2.3
NO_3-N, $kg \cdot ha^{-1}$ (0 - 30 cm)	36 ± 9	130 ± 31
NO_3-N, $kg \cdot ha^{-1}$ (30 - 60 cm)	13 ± 8	59 ± 22
NO_3-N, $kg \cdot ha^{-1}$ (60 - 120 cm)	44 ± 9	83 ± 28
Olsen-P ($mg \cdot kg^{-1}$)	14.5 ± 8.5	26.5 ± 10.2
Available-K ($mg \cdot kg^{-1}$)	333 ± 47	447 ± 45
Cation exchange capacity ($cmol_c \cdot kg^{-1}$)	NA[§]	37.1 ± 2.3

[†]Soil properties measured for 0 - 30 cm soil depth, unless stated; [‡]Values are mean ± standard deviations; [§]Not analyzed.

2.6. Measurement of N_2O Emission

During the 2012 growing season, N_2O flux measurements were conducted from the four replicate units (two undrained and two tile-drained units) in both crops. The N_2O emission rates from surface soil was measured using semi-permanent vented static PVC chamber (25.4 cm internal diameter and 10 cm height) method following the GRACEnet project protocol [31] . A polyvinyl chloride (PVC) anchor ring with beveled edge was inserted into the soil between crop rows in each sub-plot. The germinating crops, if any, inside the PVC rings were plucked out during anchor installation. Gas samples were collected at four instances for both corn (52, 78, 87, and 100 d after treatment application) and sugarbeet (35, 42, 54, and 73 d after treatment application). Gas samples were taken in between 0800 h and 1200 h of the day assuming to represent the average flux of the day. On the observation day, the height of the anchor ring above the soil surface was recorded, in order to calculate the headspace volume after chamber enclosure. A chamber was placed on the anchor and gas samples (30 mL) were collected from the chamber headspace at 0, 15, and 30 min with a graduated polypropylene syringe. The samples were then transferred to 12 mL pre-evacuated glass serum vials and transported to the laboratory for analysis. In addition, soil temperature and volumetric soil water content at the 6 cm depth, adjacent to each gas chamber, were also measured by using GS3 soil moisture-temperature sensor (Decagon Devices, Inc., Pullman, WA 99163). All the gas samples were analyzed within 24 h, using a DGA 42-Master gas chromatograph (Dani Instruments, Milan, Italy), fitted with an electron capture detector (ECD). The ECD was operated at 300°C, He carrier gas at 10 mL·min^{-1}, Hayesep N 80/100 mesh (0.32 cm diameter × 50 cm length) and Porapak D 80/100 mesh (0.32 cm diameter × 200 cm length) columns in an oven operated at 80°C. Assuming a linear increase in gas concentration, flux was calculated using the following equation:

$$F = kd\left(\frac{273}{T}\right)\left(\frac{V}{A}\right)\left(\frac{\Delta C}{\Delta t}\right)$$

where, F is the rate of gas emission (mass ha^{-1}·d^{-1}), k is unit conversion, d is gas density (g·cm^{-3}) at 273 °K, T is the air temperature (°K), V is the chamber volume (cm^3), A is soil area covered by chamber (cm^2) and ΔC/Δt is the rate of change of concentration over 15 and 30 min intervals [11].

2.7. Measurement of NH_3 Volatilization Loss

In 2013, NH_3 volatilization losses from the N fertilizers were measured from both corn and sugarbeet plots using semi-static open chambers [32] . A chamber was installed in the middle of each sub-plot in between the crop rows. The chambers were secured in an upright position on the soil surface using wire stakes, surrounded by rubber bands. Ammonia volatilization measurements were taken six times (5, 9, 19, 27, 33, and 40 d after N application), and five times (19, 22, 25, 32, and 60 d after N application) from corn and sugarbeet, respectively. On the day of measurement, the foam strips and the acid solution were collected, stored in 0.5 L mason jars containing 125 mL of 2 M KCl solution, and new traps were replaced. The sampled traps were transferred to the laboratory, where they were immediately extracted with 250 mL of 2 M KCl solution. The extracts were analyzed for NH_3 concentration using the ammonia analyzer as described above. Cumulative NH_3 volatilization loss from each sub-plot was obtained by adding NH_3produced at individual days within the sub-plot.

2.8. Measurement of Soil Water NO_3^- Concentration

Samples of the soil water at 60 cm soil depth were collected from the middle of each sugarbeet plot between rows using suction lysimeters (68 cm in length and 2.2 cm diameter; Irrometer Company, Inc., CA, USA) during the 2013 sugarbeet growing season. The suction lysimeters were installed on June 13, 2013 (15 d after treatment application) and were allowed to equilibrate for a week such that the first water sample collected (22 d after treatment application) was discarded, and not used for data analysis. Then onwards sampling was conducted for a total of 9 times during the growing season (26, 29, 33, 36, 40, 47, 54, 62, 71 d) after treatment application). Using a hand pump, a vacuum of −60 kPa was applied to the tubes and maintained for a period until the time of water sampling. Water samples inside the lysimeters were extracted using a polypropylene syringe, collected into polypropylene conical tubes, and transferred to the laboratory for analyses. In the laboratory, NO_3^- concentrations in the water samples were analyzed using the ammonia analyzer. The lysimeters were devoid of water samples in all of the tile-drained plots on 26 d, as well as, in all of the plots (undrained and tile-drained) on 47, 62, and 71 d.

2.9. Statistical Analyses

Data were analyzed separately for each year per crop using a RCBD in a split-plot arrangement with drainage as main factor and N fertilizer management as sub factor for the analysis of variance as calculated by SAS PROC GLM (version 9.3, SAS Institute Inc., Cary, NC, USA). Because of limited number of replicates, the N_2O data were pooled across the drainage management and analyzed using a RCBD with N fertilizer management as the main factor in both crops. The growing season soil inorganic N contents, N_2O fluxes, soil water NO_3^- concentration were tested separately for each sampling date. Mean separations were tested using Fisher's protected least significant difference at $P \leq 0.05$.

3. RESULTS AND DISCUSSION

3.1. Weather Conditions

Daily precipitation and mean air temperatures during the 2012 and 2013 growing seasons recorded at the research site by the Fargo NDAWN station are presented in Figure 1 [33] . In 2012, the growing season precipitation totaled 53% of the long-term (1981-2010) normal precipitation (461 mm). Consequently, the crops were under visible drought stress during the growing period. In contrast, the 2013 growing season was relatively wet compared to the normal years, but the distribution was uneven. More than half of the total season precipitation fell within the first two months of the growing season in 2013. As a result, the entire plot area was intermittently flooded during May and June of 2013. July and August were relatively dry and the crops were under drought stress. The last part of the 2013 growing season had above normal rainfall. In 2012, the growing season air temperatures were between 7°C and 25°C, which is slightly higher than the normal years. In 2013, air temperatures during the growing season ranged similar to that of the normal years, and were between 7°C and 22°C.

3.2. Soil N Availability during the Crop Growing Season

Changes in soil inorganic N availability as influenced by N-management and drainage under corn in 2012 and 2013 are presented in Figure 2. In 2012, management of N application had no influence on soil N availability throughout the growing season, regardless of drainage probably due to large inherent soil N mineralization and/or residual mineral N [6] . In fact, soil inorganic N measured during the growing season even exceeded the actual amount supplied through the N sources. In 2013, the pattern of N release in response to N management varied with drainage conditions under corn. Under the undrained condition in 2013, application of Urea134 led to a rapid buildup of inorganic N in soils during the early corn growing season (2 d after treatment application) compared to the application of urea + NP, despite both the treatments were applied at the similar N-rate. In sharp contrast under the tile-drained condition in 2013, the Urea + NP treatment delayed N release in soil until 37 d after treatment application, suggesting the potential of NP to hinder nitrification activity [9] . Corn N uptake is usually low during the early growth period [8] . And, given that the early 2013 growing season was relatively wet with a greater probability of N loss through denitrification, applying NP to urea was efficient in conserving N in soils for crop uptake during rapid crop growth. Similar to our results, previous studies have also shown that the application of NP to urea fertilizers delayed the rapid buildup of mineral N in soils compared to without NP in silty clay [9] and in silt loam [19] . Similarly, during the 2013 corn growing season, applying N fertilizer in split doses (half at planting and half at V6 corn growth stages) either resulted in lower soil N levels at early growing period (2 to 22 d) or released proportionate amount of mineral N afterwards compared to applying the entire N fertilizer at planting. Corn grain yields between these two N treatments (Urea134 and split N treatment) were not different under both drainage conditions (discussed below). Therefore, applying N in split doses could have also provided the benefit in the reduction of N loss during early growing season via denitrification, while maintaining crop yield similar to entire preplant N application [18] .

Under sugarbeet, the application of N fertilizers increased soil N levels compared to unfertilized control on 48 d under the tile-drained condition in 2012, and on 7 d under both drainage conditions in 2013 (Figure 3). The application of Urea180 extended this increment until 15 d, particularly under the tile-drained condition in 2013. The higher inorganic N levels in soils with N-fertilization over unfertilized control during the crop growing season were expected. However, application of recommended N-rate (Urea146), with or without NP, resulted in similar soil N levels as higher N rate (Urea180), regardless of drainage management. The lack of differences in soil N availabilities among the fertilizer N management practices could be attributed to the presence of large residual soil NO_3-N levels (Table 1). There were no appreciable differences in sugarbeet root yields and net sucrose concentrations among these N treatments in either year (discussed below). These results indicate that excess application of N-fertilizers should be avoided not only to attain higher gross revenue but also to reduce the risk of losses to the environment [12] [14] .

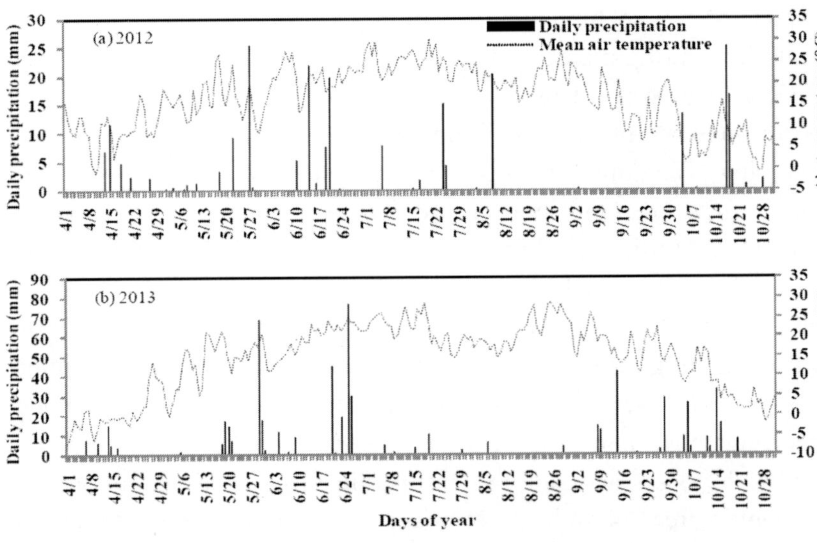

Figure 1. Daily precipitation and mean air temperature for the (a) 2012, and (b) 2013 growing seasons at the research site recorded by Fargo NDAWN station [33] .

3.3. Soil N_2O Emissions

In corn, the elevated N_2O emissions were recorded on 54 d after N application (Table 2) in response to a total of 28 mm of precipitation that occurred on June 13 - 14 (52 - 53 d after N application) (Figure 1), with the Urea224 treatment exhibiting the largest flux of 105 g N_2O-N $ha^{-1} \cdot d^{-1}$. Cumulative precipitation for the following 3 wk period totaled only 9 mm. Consequently, the emission rates measured on 78 d after N application did not exceed 10 g N_2O-N $ha^{-1}\cdot d^{-1}$ in any of the N treatments. The N_2O flux rates increased to about 27 g N_2O-N $ha^{-1}\cdot d^{-1}$ on average across N treatments on 87 d, owing to 20 mm of total precipitation received on July 24 - 25 (86 - 87 d after N application). By 100 d, the N_2O rates averaged 8 g N_2O-N $ha^{-1}\cdot d^{-1}$ on average across N treatments. Similar to corn, N_2O emissions from the sugarbeet plots were also characterized by the total amount of precipitation received during the measurement periods, with the flux values ranging between 10 to 36 g N_2O-N $ha^{-1}\cdot d^{-1}$ across N treatments (Table 2).

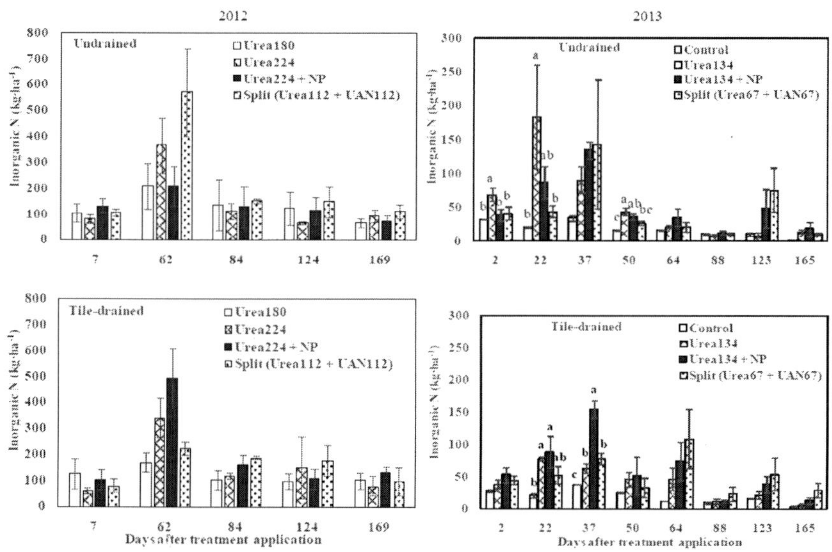

Figure 2. Inorganic N (NH_4^+ + NO_3^-) dynamics at 0 - 30 cm soil depth for N treatments under undrained and tile-drained conditions over a corn growing season in 2012 and 2013. Bars represent standard errors (n = 4). Different lower case letters within a day indicate significant difference at 0.05 level of significance.

Table 2. Soil N_2O fluxes as influenced by N management during the 2012 growing season in corn and sugarbeet.

N fertilizer	Nitrous oxide flux (g N_2O-N $ha^{-1} \cdot d^{-1}$)			
Corn	52 d	78 d	87 d	100 d
Urea180	74 ± 12	4 ± 1	25 ± 4	7 ± 1
Urea224	105 ± 69	7 ± 4	26 ± 12	11 ± 2
Urea224 + NP	74 ± 18	10 ± 2	28 ± 12	5 ± 1
Split (Urea112 + UAN112)	73 ± 17	7 ± 4	21 ± 10	6 ± 1
LSD ($P \le 0.05$)[‡]	NS[§]	NS	NS	NS
Sugarbeet	35 d	42 d	54 d	73 d
Control	10 ± 3	13 ± 4	13 ± 3	13 ± 4
Urea146	20 ± 4	15 ± 9	23 ± 5	18 ± 4
Urea146 + NP	17 ± 5	16 ± 4	24 ± 5	28 ± 13
Urea180	24 ± 7	17 ± 1	36 ± 17	29 ± 5
LSD ($P \le 0.05$)[†]	NS[§]	NS	NS	NS

[†]Values are means ± standard errors (n = 4); [‡]Least significant difference (LSD) values provided for $P \le 0.05$; [§]NS, non-significant.

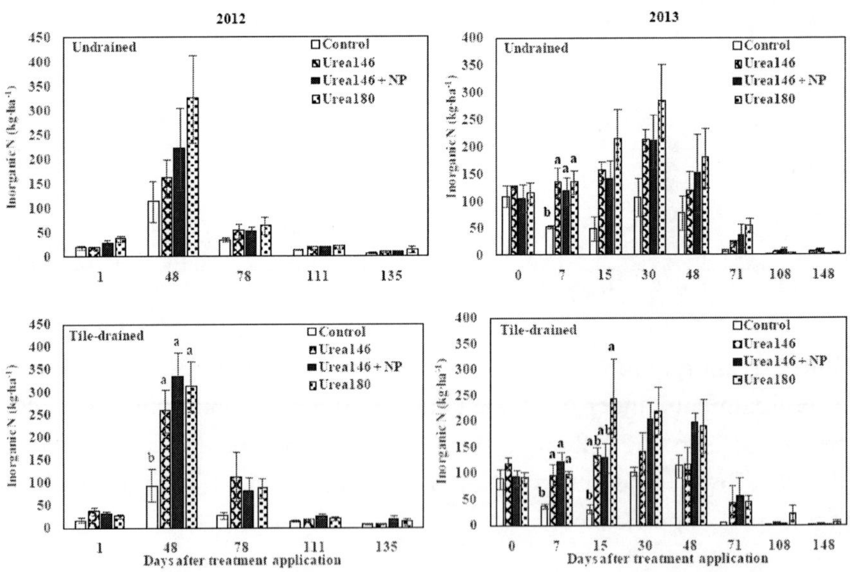

Figure 3. Inorganic N (NH_4^+ + NO_3^-) dynamics at 0 - 30 cm soil depth for N treatments under undrained and tile-drained conditions over a sugarbeet growing season in 2012 and 2013. Bars represent standard errors (n = 4). Different lower case letters within a day indicate significant difference at 0.05 level of significance.

Larger N_2O flux rates in response to the precipitation events—as a consequence of increased soil saturation— are commonly reported in the literature [10] [12] [20] . However, except for one of the flux measurement days in corn, soil N_2O fluxes remained mostly below 36 g N_2O-N $ha^{-1} \cdot d^{-1}$ across the crops, throughout the measurement period. These results are comparable to those reported for silt loam soils in Mandan, North Dakota, where peak values varied from 19 to 27 g N_2O-N $ha^{-1} \cdot d^{-1}$in spring wheat—fallow and spring wheat—safflower (Carthamus tinctorius L.)—rye (Secale cereale L.) cropping systems [34] . In contrast, daily N_2O fluxes in the present study are comparatively lower than the reported peak values of 740 g N_2O-N $ha^{-1} \cdot d^{-1}$ for silt loam soils in Indiana [19] . Poorly drained soils usually favor denitrification because of O_2 inhibition in these soils upon soil saturation [10] . In general, N_2O emissions are lower from soils with less than 60% water filled pore space (WFPS), increases slowly between 60 and 80% WFPS, and then increase more rapidly above 80% WFPS [35] . The 2012 growing season was abnormally dry (Figure 1), with the highest daily precipitation of 25 mm received for the growing season. Therefore, the dry growing period in 2012 could have restricted N_2O emissions in our study as opposed to others that have commonly reported larger spikes in N_2O production following rainfall and/or irrigation events greater than 70 mm [19] . Moreover, lower gas diffusivity and/or higher cation exchange capacity associated with fine-textured soils in the study site may have also limited N_2O production [36] .

Soil N_2O fluxes did not differ among the N management treatments on any sampling date in both corn and sugarbeet (Table 2). Studies have shown that application of mineral N fertilizers tends to cause positive and linear N_2O emissions because of greater inorganic N substrate availability [12] . Soil N availability measured at 0 - 30 cm soil depth in sugarbeet was higher with the N-fertilized treatments as compared to the control plots during the 2012 mid-growing season. However, soil N levels did not differ among the N-fertilized treatments under both corn and sugarbeet. Due to inconsistency in temporal release of mineral N from the applied fertilizers as well as soil inherent N mineralization, and due to episodic nature of N_2O emission induced by rainfall events, N_2O emissions did not vary among different N fertilizer sources during the dry growing season in 2012 [10] [37] . Besides, N_2O fluxes measured during the mid-growing season in 2012 were highly variable within the

N treatments in both crops, and consequently no significant differences were observed among the treatments [20] .

3.4. Soil NH_3 Volatilization

Cumulative losses of NH_3 did not respond to drainage management under both corn and sugarbeet in 2013 (Table 3). Approximately, 46 mm of precipitation fell within the next 5 d after N application under corn, and about 70 mm of precipitation fell on the day after N application under sugarbeet, which could have considerably incorporated the N fertilizers, regardless of drainage (Figure 1). Moreover, the N fertilizers were incorporated manually at 15 cm soil depth after N application. Therefore, soil incorporation of N fertilizer by high rainfall and/or manually after N application could have limited NH_4^+ substrate availability [38] . And considering proportionate amounts of NH_4^+ transport from the soil surface down to soil profile, NH_3 emissions from subsurface drained and undrained plots are unlikely to be different [16] .

In both corn and sugarbeet, NH_3 losses increased with N application over the unfertilized control (Table 3). Higher NH_3 losses from N-fertilized treatments are expected due to greater availability of N in the fertilized plots (Figure 2 and Figure 3). However, there were no significant differences in NH_3 losses between N sources (with vs. without NP treatments in corn) and N rates (higher vs. recommended rates in sugarbeet). In corn, across drainage, about 1.9% of applied N was lost as NH_3 from the Urea134 + NP treatment, and NH_3 volatilization from the Urea134 alone was about 1.2% of applied N. Similarly, in sugarbeet, about 0.4 and 0.5% of applied N was lost as NH_3 from Urea146 and Urea180, respectively, with no significant differences between the N-rate treatments. Studies have shown that the application of nitrification inhibitor to urea usually enhances NH_3 loss compared to untreated urea due to retention of fertilizer N in the NH_4^+ form for a prolonged period of time by the inhibitor [39] . And, NH_3 volatilization usually increases with increasing N application rate as a consequence of greater N availability [12] . However, in our study, soil N levels were similar between the N-rates throughout the sugarbeet growing season probably due to crop N uptake. And, under corn, although the availability of N tended to be greater with the application of NP during the early growing period, it apparently was not enough to significantly influence NH_3 volatilization

under wet early growing season, when most of the would have already been stabilized to the soil exchange complex [32] . Nevertheless, the amounts of NH_3 volatilization losses in our study are comparable to the previously reported values of 2.3% for surface applied urea for clay loam soil at a simulated irrigation of 16 to 19 mm of precipitation applied after 1 d of urea application [32] and 2.8% for sandy loam soils at 21.4 mm of simulated rain applied after 1 d of N application [40] . As with these studies, high precipitation (46 and 70 mm in corn and sugarbeet, respectively) received shortly after N application considerably incorporated fertilizer N into soil and restricted NH_3 volatilization [38] . Higher CEC associated with high clay soil at the research site could have

also limited the substrate required for volatilization [41] . Furthermore, the urea were incorporated into soils shortly after their application in our study, which would reduce NH_3 loss because of increased contact of urea with soil exchange complex, and where it is converted to the stable NH_4^+ form [16] .

Table 3. Cumulative NH_3 volatilization losses from N fertilizers under undrained and tile-drained conditions in corn and sugarbeet during the 2013 growing season.

N fertilizer	Cumulative NH_3 loss†		Emission factor	
	Undrained	Tile-drained	Undrained	Tile-drained
	---------------kg·ha^{-1}--------------		-----------%-----------	
Corn				
Control	1.22 ± 0.13 b	1.26 ± 0.12 b	-	-
Urea134	2.51 ± 0.39 ab	3.23 ± 0.20 a	0.96	1.47
Urea134 + NP	3.69 ± 1.27 a	3.95 ± 1.19 a	1.84	2.01
LSD ($P \leq 0.05$)‡	1.79	1.63	-	-
Sugarbeet				
Control	0.89 ± 0.11 b	0.77 ± 0.02 b	-	-
Urea146	1.44 ± 0.09 a	1.43 ± 0.14 a	0.38	0.45
Urea180	1.51 ± 0.15 a	1.75 ± 0.14 a	0.41	0.54
LSD ($P \leq 0.05$)‡	0.35	0.46	-	-

†Values are means ± standard error (n = 4); Different lowercase letters within a column indicate significant difference at 0.05 level of significance.
‡Least significant difference (LSD) value provided for $P \leq 0.05$.

3.5. Soil Water NO_3^- Concentration

Soil water NO_3^- concentration ($mg{\cdot}L^{-1}$) measured at the 60 cm soil depth during the 2013 sugarbeet growing season for the N treatments under undrained and subsurface drained conditions are presented in Figure 4. The concentrations of NO_3^- at 60 cm depth was highly variable. However, the concentrations of NO_3^- were slightly lower under subsurface drained than undrained condition across all the N treatments throughout the measurement period. Under the undrained condition, peak concentrations of NO_3^- in Urea146, Urea146 + NP, and Urea180 were 96, 75, and 85 $mg{\cdot}L^{-1}$, respectively, whereas the corresponding peak concentrations for these treatments under the tile-drained condition were 53, 69, and 63 $mg{\cdot}L^{-1}$, respectively. A consistently lower NO_3^- concentration with tile drainage could in part be due to greater N uptake by sugarbeet crops under tile drained condition. Sugarbeet N demand is usually high from the early growth period until canopy growth phase for the development of above-ground plant parts, during which maximum N is assimilated [42] . And, the measurement of leaf chlorophyll conducted at 54 d revealed that, across the N treatments, the spad values were slightly higher for the tile drained sugarbeet plots compared to undrained plots (data not presented).

Soil NO_3^- concentrations varied among the N treatments with drainage. Nitrate concentration generally increased with N fertilization over the control irrespective of drainage management, however. The lower NO_3^- levels with the control treatments were expected. Under the tile-drained condition, application of NP accumulated significantly greater soil water NO_3^- levels than the control, while the urea treatments without NP were only slightly greater than control. Under the undrained condition, the soil water concentration appeared not to depend on NP but N fertilization. In the present study, the extraction of soil water for the determination of NO_3^- concentration began only after 29 d following N application due to untrafficable soil condition at the study site. Despite the delayed measurements, under the tile-drained condition, the NO_3^- concentration tended to be slightly higher with the application of Urea146 + NP than without NP treatments (Urea146 and Urea180). Since nearly a third of the total growing season precipitation fell within the initial 30 d (Figure 1), considerable losses of N were deemed possible through denitrification without NP application during these initial 4 wk period [9] [19] . These results suggest the efficacy of nitrification inhibitor NP to conserve more NO_3^- in soils underthe tile-drained condition, which otherwise could have potentially lost through denitrification without the application of NP [10] .

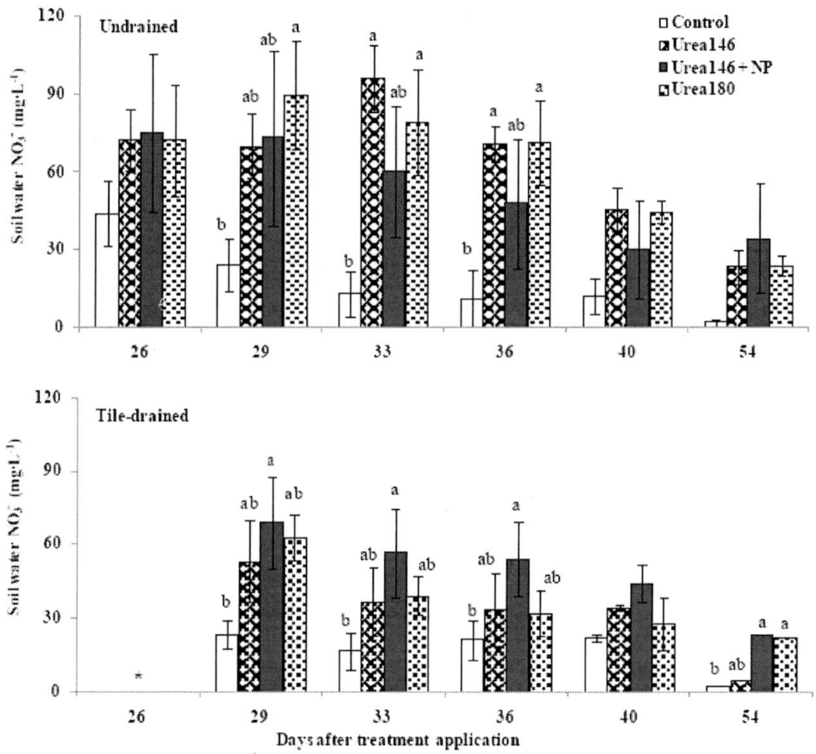

Figure 4. Soil water NO_3^- concentration (mg·L^{-1}) measured during the 2013 sugarbeet growing season as influenced by N management under undrained and drained conditions. Bars represent standard errors (n = 4). Different lower case letter within a day indicate significant difference at 0.05 level of significance. *Not available for the day.

3.6. Crop Yields

Tile drainage had no influence on crop yields and quality parameters during the 2012 and 2013 growing seasons (Table 4 and Table 5). Several studies have indicated significant crop yield improvements under tile drainage compared with undrained conditions due to improvement in soil environment and/or N availability for crops with drainage [3] [4] . However, crop yield response to drainage management can vary with the amount and the pattern of precipitation received during the growing season [10] . And, studies have confirmed that water stress during reproductive growth stages of crops can dramatically reduce yields [43] . In the current study, the crops experienced drought conditions during the growing periods in both years. Therefore, soil water deficit was likely the most limiting factor affecting crop yields due to abnormally dry growing conditions. In addition, water stress could have also lowered N demand in the crops. Consequently, over the two study years, the expected benefit of tile drainage to improve crop yields was likely minimized, regardless of N management.

Table 4. Corn grain yield as affected by N management under undrained and tile-drained conditions in 2012 and 2013.

N fertilizer	Grain yield[†]	
	Undrained	Tile-drained
	---Mg·ha^{-1}---	
2012		
Urea180	7.46 ± 0.81 a	8.10 ± 0.82 a
Urea224	6.84 ± 0.72 a	7.53 ± 0.92 a
Urea224 + NP	7.56 ± 1.09 a	6.42 ± 0.49 a
Split (Urea112 + UAN112)	7.71 ± 1.01 a	7.38 ± 0.37 a
LSD (P ≤ 0.05)[‡]	NS[§]	NS
2013		
Control	5.97 ± 0.41 b	6.01 ± 0.27 b
Urea134	8.79 ± 0.37 a	8.33 ± 0.57 a
Urea134 + NP	8.65 ± 0.59 a	8.59 ± 0.68 a
Split (Urea67 + UAN67)	7.42 ± 0.19 a	7.84 ± 0.57 a
LSD (P ≤ 0.05)[‡]	1.43	0.86

[†]Values are means ± standard error (n = 4); Different lowercase letters within a column within each year indicate significant difference at 0.05 level of significance; [‡]Least significant difference (LSD) values provided for P ≤ 0.05; [§]NS, non-significant.

Table 5. Sugarbeet root yield and quality parameters as affected by N management under undrained and tile-drained conditions in 2012 and 2013.

N fertilizer	Root Yield[†]		Sucrose loss to molasses[†]		Net sugar[†]	
	Undrained	Tile-drained	Undrained	Tile-drained	Undrained	Tile-drained
2012	----------Mg·ha^{-1}----------		----------------------------------%--------------------------------			
Control	43.8 ± 1.9 a	51.3 ± 5.4 a	1.62 ± 0.04 a	1.68 ± 0.05 b	16.6 ± 0.4 a	16.9 ± 0.6 a
Urea146	47.0 ± 4.7 a	48.6 ± 4.0 a	1.74 ± 0.09 a	1.86 ± 0.12 ab	16.9 ± 0.8 a	16.6 ± 0.2 a
Urea146 + NP	44.7 ± 3.1 a	45.5 ± 5.3 a	1.75 ± 0.07 a	1.95 ± 0.12 a	17.1 ± 0.4 a	16.5 ± 0.2 a
Urea180	45.6 ± 1.3 a	48.8 ± 5.8 a	1.89 ± 0.06 a	2.03 ± 0.08 a	16.1 ± 0.3 a	15.6 ± 0.2 a
LSD (P ≤ 0.05)[‡]	NS[§]	NS	NS	0.21	NS	NS
2013						
Control	39.4 ± 1.6 a	36.7 ± 2.2 a	1.54 ± 0.14 a	1.69 ± 0.20 a	14.5 ± 0.4 a	14.4 ± 0.4 a
Urea146	40.1 ± 0.5 a	41.0 ± 1.1 a	1.58 ± 0.05 a	1.81 ± 0.12 a	14.3 ± 0.5 a	13.5 ± 0.6 a
Urea146 + NP	39.6 ± 0.6 a	37.0 ± 1.7 a	1.74 ± 0.08 a	1.67 ± 0.08 a	13.8 ± 1.6 a	14.3 ± 0.2 a
Urea180	36.7 ± 1.0 a	35.8 ± 3.8 a	1.73 ± 0.10 a	1.68 ± 0.12 a	14.0 ± 0.1 a	13.9 ± 0.5 a
LSD (P ≤ 0.05)[‡]	NS[§]	NS	NS	NS	NS	NS

[†]Values are means ± standard error (n = 4); Different lowercase letters within a column within each year indicate significant difference at 0.05 level of significance; [‡]Least significant difference (LSD) values provided for P < 0.05; [§]NS, non-significant.

In 2012, corn yields averaged 7.39 and 7.35 Mg·ha^{-1} under undrained and tile-drained conditions, respectively, with no significant differences among the N fertilizer management sources (Table 4). Lack of yield responses to fertilizer N management practices could be attributed in part to large soil residual N and/or organic N mineralization during the growing seasons [37] . In fact, soil N levels measured during the 2012 growing season even exceeded the amount supplied through N fertilizers (Figure 2). Moreover, soil N availabilities among the N fertilizer sources and rates were similar throughout the growing season under both drainage conditions. Given that soil N availability from the N sources were similar, it can be safely presumed that N uptake by corn plants were similar for the N sources, which likely limited yield differences among them. Additionally, the 2012 growing season was abnormally dry and the movement of N into the active root zone might have also been hindered within the dry topsoil that consequently may have led to poor root N uptake [44] . Nevertheless, in 2013, corn yields increased with the application of fertilizer-N over the unfertilized control on average by 2.32 Mg·ha^{-1} (27.9% increment) under the undrained condition and by 2.24 Mg·ha^{-1} (27.2% increment) under the tile-drained condition. But, the yields were similar among the N-fertilized treatments under both undrained and drained conditions, and averaged 8.29 $Mg \cdot ha^{-1}$ and 8.25 $Mg \cdot ha^{-1}$, respectively. These results suggest the necessity of N application regardless of N sources in order to optimize crop yields [18] . However, the lack of yield differences among the N-fertilized treatments, regardless of N source and N rate could be attributed to drought stress during the dry mid-growing period that likely reduced the overall corn yield potential, obscuring the effects of N management [12] [44] . Our results are in line with previous studies that have also reported no responses to the application of N sources and rates containing urease inhibitor, nitrification inhibitor, or slow release polymer coated urea under abnormally dry growing periods [10] [12] [45] . The response of N fertilizers on crop yields are usually substantial in coarse-textured soils with higher crop N demands under irrigated condition, whereas the response of N fertilizers may be limited in fine-textured soils having low crop N demand with limited soil water availability [46] .

Similar to corn, N management had no influence on sugarbeet root yield and sucrose concentration in both years (Table 5). Across N treatments, in 2012, sugarbeet root yield averaged 45.3 and 48.6 $Mg \cdot ha^{-1}$, with net sucrose concentrations of 16.7 and 16.4% under undrained and tile-drained conditions, respectively. In 2013, the average sugarbeet root yields were 39.0 and 37.6 $Mg \cdot ha^{-1}$, and sucrose concentrations were 14.2 and 14.0% under undrained and tile-drained conditions, respectively. Soil residual NO_3-N contents measured in the sugarbeet plots were 93 and 272 $kg \cdot ha^{-1}$ (at 0 - 120 cm soil profile) for 2012 and 2013, respectively (Table 1). Such high residual N levels could have significantly contributed crop N needs, and consequently no response to the added N through fertilizers was observed. Instead, the application of N increased impurity (SLM %) in beet roots, particularly under the tile-drained condition in 2012 (Table 5). Sugarbeet root impurity increased by 13.8% with Urea146 + NP and by 17.2% with Urea180 compared to unfertilized control under the tile-drained condition in 2012. The reduction in sugarbeet root quality with N application could be due to greater crop N uptake with Urea146 + NP and Urea180 treatments because soil N levels measured during the mid-growing season were significantly higher with these treatments as compared to unfertilized control (Figure 3). Previous studies have confirmed that greater soil N availability than required usually lowers purity indexes with depressed sucrose concentrations [47] . Excess N fertilization can stimulate sugarbeet top growth beyond the point where maximum root yields are attained and thereby direct photosynthates into regenerating canopy rather than into the root storage [48].

4. CONCLUSION

Our experiment showed the need for long term studies of subsurface drainage and N management on crop yields and quality, and N availability and N losses in the RRV. Although contrasting weather patterns occurred during the 2012 and 2013 growing seasons, our study emphasized that across intense wetting and drying cycles, subsurface drainage and N management influenced N availability more than crop yield in the silty clay soil. Prolong periods of soil saturation due to poor internal drainage has typically been the greatest limiting factor affecting crop yields in this soil. And, contrary to the norm, soil water deficit was likely the most limiting factor affecting crop yields over the two study years, due to abnormally dry conditions experienced during the growing seasons. Additionally, inherent soil N mineralization appears to be an important factor controlling crop yields and N losses, and therefore assessment of soil mineral N during crop growth period appears to be important to improve our knowledge on N availability for crops. Application of higher than recommended N rates had no effect on yields, but was only associated with more daily soil N_2O emissions. Conversely, the application of recommended N fertilizer rates along with nitrification inhibitor still holds as a viable N management strategy to increase N availability to crops without compromising yields. Also, nitrification inhibitor might have the potential to conserve soil water NO_3^- concentrations at 60 cm depth, particularly under the tile-drained condition. Apart from the high cation exchange associated with the soil under investigation, incorporation of fertilizer-N into soils mechanically and/or rainfall can considerably restrict NH_3 volatilization from the soil, with a possibility of up to 1.9% loss from the applied-N. The research results may provide important information to growers considering suitable N management under subsurface drainage systems within the RRV.

REFERENCES

1. US Department of Agriculture—Natural Resources Conservation Service (2014) Soil Survey Office, Fargo, ND.

2. Kandel, H.J., Brodshaug, J.A., Steele, D.D., Ransom, J.K., Desutter, T.M. and Sands, G.R. (2013) Subsurface Drainage Effects on Soil Penetration Resistance and Water Table Depth on a Clay Soil in the Red River of the North Valley, USA. Agricultural Engineering International: CGIR Journal, 15, 1-10.

3. Kladivko, R.L., Willowghby, G.L. and Santini, J.B. (2005) Corn Growth and Yield Response to Subsurface Drain Spacing on Clermont Silt Loam Soil. Agronomy Journal, 97, 1419-1428.

4. Jin, C.X., Sands, G.R., Kandel, H.J., Wiersma, J.H. and Hansen, B.J. (2008) Influence of Subsurface Drainage on Soil Temperature in a Cold Climate. Journal of Irrigation and Drainage Engineering, 34, 83-88.

5. Agehara, S. and Warncke, D.D. (2005) Soil Moisture and Temperature Effects on Nitrogen Release from Organic Nitrogen Sources. Soil Science Society of America Journal, 69, 1844-1855.

6. Rochette, P., Tremblay, N., Fallon, E., Angers, D.A., Chantigny, M.H., MacDonald, J.D., Bertrand, N. and Parent, L.E. (2010) N2O Emissions from an Irrigated and Non-Irrigated Organic Soil in Eastern Canada as Influenced by N Fertilizer Addition. European Journal of Soil Science, 61, 186-196.

7. Dessureault-Rompre, J.B., Zebarth, B.J., Burton, D.L., Gregorich, E.G., Goyer, C., Georgallas, A. and Grant, C.A. (2013) Are Soil Mineralizable Nitrogen Replenished during the Growing Season in Agricultural Soil? Soil Science Society of America Journal, 77, 512-524.

8. Zebarth, B.J. and Paul, J.W. (1997) Growing Season Nitrogen Dynamics in Manured Soils in South Coastal British Columbia: Implications for a Soil Nitrate Test for Silage Corn. Canadian Journal of Soil Science, 77, 67-76.

9. Awale, R. and Chatterjee, A. (2015) Soil Moisture Controls the Denitrification Loss of Urea-Nitrogen from Silty Clay Soil. Communications in Soil Science and Plant Analysis, 46, 2100-2110.

10. Nash, P., Nelson, K. and Motavalli, P. (2015) Reducing Nitrogen Loss with Managed Drainage and Polymer-Coated Urea. Journal of Environmental Quality, 44, 256-264.

11. Ginting, D., Kessavalou, A., Eghball, B. and Doran, J.W. (2003) Greenhouse Gas Emissions and Soil Indicators Four Years after Manure and Compost Applications. Journal of Environmental Quality, 32, 23-32.

12. Thapa, R., Chatterjee, A., Johnson, J.M.F. and Awale, R. (2015) Stabilized Nitrogen Fertilizers and Application Rate Influence Nitrogen Losses under Rainfed Spring Wheat. Agronomy Journal, 107, 1-10.

13. Drury, C.F., Tan, C.S., Reynolds, W.D., Welacky, T.W., Oloya, T.O. and Gaynor, J.D. (2009) Managing Tile Drainage, Subirrigation, and Nitrogen Fertilization to Enhance Crop Yields and Reduce Nitrate Loss. Journal of Environmental Quality, 38, 1193-1204.

14. Randall, G.W. and Vetsch, J.A. (2005) Nitrate Losses in Subsurface Drainage from a Corn-Soybean Rotation as Affected by Fall and Spring Application of Nitrogen and Nitrapyrin. Journal of Environmental Quality, 34, 590-597.

15. Singh, U., Sanabria, J., Austin, E.R. and Agyin-Birikorang, S. (2011) Nitrogen Transformation, Ammonia Volatilization Loss, Nitrate Leaching in Organically Enhanced Nitrogen Fertilizers Relative to Urea. Soil Science Society of America Journal, 76, 1842-1854.

16. Norman, R.J., Wilson Jr., C.E. and Slaton, N.A. (2003) Soil Fertilization and Mineral Nutrition in US Mechanized Rice Culture. In: Smith, W. and Dilday, R.H., Eds., Rice: Origin, History, Technology, and Production, John Wiley & Sons Inc., Hoboken, 331-411.

17. Al-Kanani, T., MacKenzie, A.F. and Barthakur, N.N. (1991) Soil Water and Ammonia Volatilization Relationships with Surface-

Applied Nitrogen Fertilizer Solutions. Soil Science Society of America Journal, 55, 1761-1766.

18. Maharjan, B., Venterea, R.T. and Rosen, C. (2014) Fertilizer and Irrigation Management Effects on Nitrous Oxide Emissions and Nitrate Leaching. Agronomy Journal, 106, 703-714.

19. Omonode, R.A. and Vyn, T.J. (2013) Nitrification Kinetics and Nitrous Oxide Emissions When Nitrapyrin Is Coapplied with Urea-Ammonium Nitrate. Agronomy Journal, 105, 1475-1486.

20. Parkin, T.B. and Hatfield, J.L. (2013) Influence of Nitrapyrin on N2O Losses from Soil Receiving Anhydrous Ammonia. Agriculture, Ecosystems and Environment, 136, 81-86.

21. Soil Survey Staff (2014) Keys to Soil Taxonomy. 12th Edition, USDA-Natural Resources Conservation Service, Washington DC.

22. Franzen, D.W. (2010) North Dakota Fertilizer Recommendation Tables and Equations. Publication No. SF-882 (Revised), NDSU Extension Service, North Dakota State University, Fargo. https://www.ndsu.edu/fileadmin/soils/pdfs/sf882.pdf

23. Maynard, D.G., Kalra, Y.P. and Crumbaugh, J.A. (2008) Nitrate and Exchangeable Ammonium Nitrogen. In: Carter, M.R. and Gregorich, E.G., Eds., Soil Sampling and Methods of Analysis, Canadian Society of Soil Science, CRC Press, Boca Raton, 71-80.

24. Blake, G.R. and Hartge, K.H. (1986) Bulk Density. In: Klute, A., Ed., Methods of Soil Analysis, Part 1, ASA and SSSA, Madison, 363-367.

25. Combs, S.M. and Nathan, M.V. (1998) Soil Organic Matter. In: Brown, J.R., Ed., Recommended Chemical Soil Test Procedures for the North Central Region, NCR Publication No. 221, Missouri Agricultural Experiment Station SB 1001, University of Missouri, Columbia, 53-58.

26. Thomas, G.W. (1996) Soil pH and Acidity. In: Sparks, D.L., Ed., Methods of Soil Analysis, Part 3, ASA, SSSA, Madison, 475-490.

27. Chapman, H.D. (1965) Cation-Exchange Capacity. In: Black, C.A., Ed., Methods of Soil Analysis, Part 2: Chemical and Microbiological Properties, ASA, Madison, 891-900.

28. Elliott, E.T., Heil, J.W., Kelly, E.F. and Monger, H.C. (1999) Soil Structural and Other Physical Properties. In: Robertson, G.P., et al., Eds., Standard Soil Methods for Long Term Ecological Research, Oxford University Press, Inc., New York, 74-88.

29. Frank, K., Beegle, D and Denning, J. (1999) Phosphorus. In: Brown, J.R., Ed., Recommended Chemical Soil Test Procedures for the North Central Region, NCR Publication No. 221, Missouri Agricultural Experiment Station SB 1001, University of Missouri, Columbia, 21-30.

30. Warncke, D. and Brown, J.R. (1998) Potassium and Other Basic Cations. In: Brown, J.R., Ed., Recommended Chemical Soil Test Procedures for the North Central Region, NCR Publication No. 221, Missouri Agricultural Experiment Station SB 1001, University of Missouri, Columbia, 31-33.

31. Parkin, T.B. and Venterea, R.T. (2010) Chamber-Based Trace Gas Flux Measurements. In: Follet, R.F., Ed., Sampling Protocols, USDA-ARS, Washington DC, 3.1-3.39. http://www.ars.usda.gov/research/GRACEnet

32. Jantalia, C.P., Halvorson, A.D., Follett, R.F., Alves, B.J.R., Polidoro, J.C. and Urquiaga, S. (2012) Nitrogen Source Effects on Ammonia Volatilization as Measured with Semi-Static Chambers. Agronomy Journal, 104, 1595-1603.

33. NDAWN (2015) North Dakota Agricultural Weather Network Center. http://ndawn.ndsu.nodak.edu

34. Liebig, M.A., Tanaka, D.L. and Gross, J.R. (2010) Fallow Effects on Soil Carbon and Greenhouse Gas Flux in Central North Dakota. Soil Science Society of America Journal, 74, 358-365.

35. Bateman, E.J. and Baggs, E.M. (2005) Contribution of Nitrification and Denitrification to N2O Emissions from Soils at Different Water-Filled Pore Space. Biology and Fertility of Soils, 41, 379-388.

36. Gu, J., Nicoullaud, B., Rochette, P., Grossel, A., Henault, C., Cellier, P. and Richard, G. (2013) A Regional Experiment Suggests That Soil Texture Is a Major Control of N2O Emissions from Tile-Drained Winter Wheat Fields during the Fertilization Period. Soil Biology and Biochemistry, 60, 134-141.

37. Dell, C.J., Han, K, Bryant, R.B. and Schmidt, J.P. (2014) Nitrous Oxide Emissions with Enhanced Efficiency Nitrogen Fertilizers in a Rainfed System. Agronomy Journal, 106, 723-731.

38. Rochette, P., Angers, D.A., Chantigny, M.H., Gasser, M., MacDonald, J.D., Pelster, D.E. and Bertrand, N. (2013) Ammonia Volatilization and Nitrogen Retention: How Deep to Incorporate Urea? Journal of Environmental Quality, 42, 1635-1642.

39. Gioacchini, P., Nastri, A., Marzadori, C., Giovannini, C., Antisari, L.V. and Gessa, C. (2002) Influence of Urease and Nitrification Inhibitors on N losses from Soils Fertilized with Urea. Biology and Fertility of Soils, 36, 129-135.

40. Holcomb III, J.C., Sullivan, D.M., Horneck, D.A. and Clough, G.H. (2011) Effect of Irrigation Rate on Ammonia Volatilization. Soil Science Society of America Journal, 75, 2341-2347.

41. Griggs, B.R., Norman, R.J., Wilson Jr., C.E. and Slaton, N.A. (2007) Ammonia Volatilization and Nitrogen Uptake for Conventional and Conservation Tilled Dry-Seeded, Delayed-Flood Rice. Soil Science Society of America Journal, 71, 745-751.

42. Martin, S.S. (2001) Growing Sugarbeet to Maximize Sucrose Yield. In: Wilson, R.G., Ed., Sugarbeet Production Guide, Bulletin EC01-156, Nebraska Coop. Ext., University of Nebraska, Lincoln, NE, 3-8.

43. Cakir, R. (2004) Effect of Water Stress at Different Development Stages on Vegetative and Reproductive Growth of Corn. Field Crops Research, 89, 1-16.

44. Ray, J.D., Heatherly, L.G. and Fritschi, F.B. (2005) Influence of Large Amounts of Nitrogen on Nonirrigated and Irrigated Soybean. Crop Science, 46, 52-60.

45. Mckenzie, R.H., Middleton, A.B., Pfiffner, P.G. and Bremer, E. (2010) Evaluation of Polymer-Coated Urea and Urea Inhibitor for Winter Wheat in Southern Alberta. Agronomy Journal, 102, 1210-1216.

46. Abalos, D., Jeffery, S., Sanz-Cobena, A., Guardia, G. and Vallejo, A. (2014) Meta-Analysis of the Effect of Urease and Nitrification Inhibitors on Crop Productivity and Nitrogen Use Efficiency. Agriculture, Ecosystems and Environment, 189, 136-144.

47. Halvorson, A.D. and Hartman, G.P. (1974) Long-Term Nitrogen Rates and Sources Influence Sugarbeet Yield and Quality. Agronomy Journal, 67, 389-393.

48. Anderson, F.N. and Peterson, G.A. (1988) Effect of Incremental Nitrogen Application on Sucrose Yield of Sugarbeet. Agronomy Journal, 80, 709-712.

Index